THE GREENHOUSE EFFECT

A LEGACY

A NOVEL OF LIVING WITH CLIMATE CHANGE

AND

A SCIENTIST'S BRIEF ON GLOBAL WARMING

by

Alex Cook

For the Gilpin Library
Alex Cook
Alex R. Burnett

FIC
COOK

First published by Dog Ear Publishing
4010 W. 86th Street, Ste H
Indianapolis, IN 46268
www.dogearpublishing.net

dog ear
PUBLISHING

ISBN: 978-159858-348-9

This book is printed on acid-free paper.

Printed in the United States of America

ACKNOWLEDGEMENTS

The advice and assistance of Barbara Perrin of Ibis Publication Consulting is gratefully acknowledged. My state of mind was set at ease about this enterprise.

The prompts in Microsoft Word on spelling and sentence structure were invaluable; those that were ignored are entirely my fault. Finally, though, I'm indebted to daughter Pamela for expertise on matters of grammar.

Encouragement from neighbors, family members, and other friends helped move this project on global warming to completion.

Cover photo by the author.
Photo of the author by daughter Marcia Montague, (1952-2003).

AUTHOR'S NOTE:

This story about Calvin Carpenter, Kathy Martin, and their Kansas neighbors is fictional. All names, characters, places, and events are either completely fictitious or are used fictitiously. In particular, descriptions of weather, climate, and related human effects are not historical, nor are these predictions. Experts will be hard pressed to deny these possibilities however. On the other hand, the website contributions describing the fundamentals of atmospheric science and the potential consequences of climate change by the fictional Chris Baldwin are solid defensible science. The blog messages by the fictional Carlos Martinez continue with more proactive demands to recognize and correct the dangerous practice of reliance on fossil fuels. The message that I wish to convey is that climate change is an inevitable response to the influence of increasing human populations. It deserves our understanding and attention. I, alone, am responsible for these discussions on global warming. Any institutions or government agencies mentioned are not involved in any way.

Alex Cook

Rollinsville, CO

THE GREENHOUSE EFFECT
A LEGACY
CONTENTS

Chapter III. Heavy Weather

Chapter IV. Recovery

Chapter V. Commitments

INTRODUCTION

As the sun crossed the equator on March 21, the solar intensity on the upper atmosphere of the earth was essentially the same as that for the past several hundred years. But the earth's infrared blanket was slightly thicker; the air and the surfaces of the land and ocean were perceptibly warmer. As the earth continued on its nearly circular orbit towards the summer solstice in June, the days grew longer and the midday sun at 23 degrees north latitude approached the zenith.

Calvin Carpenter, recently retired physical science instructor at Central Plains Community College, begins to settle in to a quiet life on the small family farm on Raccoon Creek, a tributary of the Arkansas. He is resigned to grow old alone. Still, in retirement he feels an obligation to transmit his scientific knowledge of the atmospheric environment to his neighbors and friends. Calvin's neighboring farm resident is Kathy Martin, a widow winding down from an active career in professional music. Calvin and Kathy have each learned to appreciate the comfortable life and the quiet beauty of the riverside environment.

But mankind has become a major player in Nature. Increasing populations and technological advances have moved the planet into the initial phases of global warming. Earth's climate continues in its inexorable advance to a new regime. Droughts will occur, glaciers will melt, and sea levels will rise. The earth's atmosphere will be forced to deal with an increasing problem of energy transport from the tropics to high latitudes.

The consequences promise to be a major perturbation to the lives of Calvin and his neighbors in southeast Kansas. They will be forced to adapt to a less benign climate.

Chapter One

CALVIN CARPENTER
AND CLIMATE CHANGE

Lunchtime Presentation

A t 1:10 PM on Tuesday April 23 in the Commons Room of the Community Center at Lynn, Kansas, members of the Chamber of Commerce were pushing aside their dessert plates and receiving refills of Starbucks coffee. The apple pie a la mode from Aunt Ida's Ice Cream Parlor had been uniformly demolished and the young ladies of Lynn Junior Class, filling T-shirts with breasts swelling with reminders of City Delicatessen and backs advertising Starbucks Coffee, were returning to the kitchen.

Mayor Alvin Clarkson in a three-piece suit, resplendent with watch fob and gold chain, stood with some effort, then began to signal attention with spoon and empty water glass. "Having disposed of our trivial items of business prior to lunch, we now have more exciting and weightier topics to consider. We are privileged to have as our guest today Dr. Calvin Carpenter, Professor Emeritus of Physical Science from Central Plains Community College up the road a piece at Jamestown. Cal has been our neighbor on that farm down on Raccoon Creek off and on for many years. Benjamin and Mary settled there at the turn of the century; they are buried in the old cemetery just east of town. Cal leased the farm to Swenson's for a few years as you know, then came back to live in the fine old farmhouse a few months ago after his retirement from teaching at the college.

Cal earned his Ph.D. in Physics at the big campus up north and did some teaching and research there for a few years before settling in at Central Plains. I understand his old man received training as a weatherman for the U.S. Air Corps and Cal got interested in the subject. He says his specialty now is Physics of the Atmosphere. I have asked him to come today and tell us what he knows of the Greenhouse Effect and Global Warming. So Professor, it's all yours!"

Professor Carpenter stood slowly to face the Mayor, "Are you sure you want to listen to all I know about that subject? That may take at least 10 minutes!" Polite smiles and a few chuckles from those at nearby tables. But they were confident of facts and predictions from the horse's mouth.

Then, facing his audience of several dozen local business men and a few farmers, he cautioned, "Seriously, I need to make clear that I am by no means a recognized expert and world authority on this complex subject. However, I have given some attention and time over the years to some related atmospheric measurements, some of which have become standard procedures. And I have become exposed to much of the recent theory relating to climate change. I will give a brief broad-brush treatment of the subject to you today."

Calvin had evidently been a lean, athletic, 6-foot college lecture personality in his earlier years. Today, he appeared slightly stooped, but still wearing his dark rimmed glasses of yesteryear. Blonde hair, somewhat thin on top, flowed into a scraggly ponytail. Sideburns were neatly trimmed however; the pale handlebar mustache had received similar care. His casual removal of the brown tweed jacket revealed narrow blue suspenders attached to dark woolen dress slacks. "Now it has been my experience that an after lunch lecture often results in nap time for about half my audience. Consequently, I refuse to darken the room for boring slides; you'll have to sleep with eyes open." There were a couple of very pretentious loud groans at this.

"Now let me tell you of some of the controversies about global warming and the factual bases for much of the concern. Then, before I've told you more than you ever wanted to know about tomorrow's climate, I'll ask for questions."

Calvin introduced the concept of the Greenhouse Effect and its role in sustaining Earth's temperatures suitable for human habitation. He then gave some concentrated attention to the 50-year database on atmospheric carbon dioxide from Mauna Loa, Hawaii, and the attendant concern about long term global warming resulting from the burning of fossil fuels. He continued with warnings of the observations of glacial retreat and Arctic warming, followed by projections of sea level rise in the wake of possible melting of the Greenland and Antarctic ice caps. Finally, he described the long but noisy average temperature records and the projected computer predictions. "Now, I'm sure you must have some specific questions. Ask away!"

Jesse Jones, farmer, was ready with a quick question. "I've seen predictions of an average temperature increase of a degree or so in the next few years. Now I had the impression that global warming was a slow process, not something that I needed to plan my crops for. And a degree or so doesn't worry me; but at that rate, it's something that my grandchildren might have a big problem with. Is this what they call a clear and present danger?"

"There are data from glacier ice cores that show significant past warmings in a matter of decades. Our recent temperature changes suggest we might be following that pattern. In any case, it's important to think about it now!"

Bank president Basil Jefferson had made an effort to dress casually with a checked sport coat and a bow tie. He still had the somewhat pompous posture of wealth however. "You've said that the temperature record is noisy. Isn't there some doubt about the dependability of those measurements?"

"Yes, there are measurement uncertainties and the analysis is complicated. And there are natural variations of course—we see that locally from year to year. It's the long-term trend in the temperature measurements that has the most significance. The correlation of average global temperature with the very accurate measurements of atmospheric carbon dioxide of the past 50 years is quite meaningful. It is the connection of the CO_2 greenhouse gas increase to our burning of fossil fuels that is dangerous and worrisome."

Russell Harding owned the John Deere farm implement store. "How can carbon dioxide make such a difference? You just told us that the Greenhouse Effect has kept the earth warm for most of its history, maybe millions of years. Things like water must be a lot more important. And why blame fossil fuels?"

"Yes, water is an extremely important greenhouse gas. However, CO_2 behaves in similar fashion and it is the one that mankind has increased markedly in recent years. Burning any fuel containing carbon will produce carbon dioxide of course. Even major forest fires do produce a minor burp of CO_2, but it is the steady increase in CO_2 that is blamed on burning of fossil fuels."

Harding replied, "But our society will go back to the dark ages if we don't burn coal and oil. These environmentalists are just trying to scare us. Now our government leaders have made a proper decision that things like the Kyoto deal are bad for our economy! And some of those important people tell us it's all a fraud!"

The pharmacist from the Corner Drug interrupted. "Dave Crause here, and I guess I must be an environmentalist because this air pollution scares me. I believe these international agreements are a good thing. Didn't that Montreal protocol save us from a real bad ultraviolet problem?"

"I'm not so convinced of that," was Harding's reply. "I remember reading that our Maker designed the Earth as a living thing that could respond to all these problems. Isn't that right, Pastor Swenson?"

The Baptist pastor apologetically answered, "With all due respect to non-believers and one or two Moslems in the audience, I tell my flock that if we live a Christian life our God will watch over us. We elect our leaders based on these moral values. There is no doubt in my mind that their decisions will be best for our society."

It was Dave Crause who jumped to his feet. "I protest! Our scientists are not evil people. We should not ignore their recommendations!"

The Mayor was not eager for a religious debate. "Whoa now! Let's get back to our guest. I suspect he has something to say about ozone."

"Thanks Mayor. I do indeed! The ozone problem was very real and we had a narrow escape. If we had procrastinated for a few more years, there would have been a worldwide catastrophe. Mad dogs and Englishmen aside, none of us would have been out in the ultraviolet of the midday sun! Now the atmosphere is slowly healing. Even though we have stopped putting those chlorine compounds in the air from our leaky air conditioners, that old reservoir has a lifetime of decades. Those ozone holes will not disappear overnight. And a word of caution. The lifetime of CO_2 in the environment is about 100 years. If we are making a mistake there, it will be even more difficult to correct!"

But Harding would not be silenced. "I've been told that if we keep a healthy economy that our technology advances will solve those problems. We should expect that sending men to Mars will work miracles like it did with computers to go to the moon. There must be some way to take CO_2 out of the air!"

"Well now! I have an obvious solution." Jerry Amundsen operated the refrigeration system at the meat packing plant. "You can liquefy carbon dioxide, even make dry ice." Then he chuckled, "But there's one little drawback! It takes a lot of electric power to operate those compressors; that coal burning power plant up north will stay ahead of us on CO_2! There's a better solution—plant trees!"

Calvin added, "That's correct! The natural growth process sequesters carbon in the wood. We should harvest our forest in sustainable fashion—plant new trees. Slash and burn farming is no good. That and natural fires put CO_2 into the air and the trees are not replaced. New forests would be helpful, but we will also have to slow down on burning coal and oil. Energy sources like solar, wind, and nuclear need more attention."

"These alternative energy sources are just an expensive distraction. Yes, you've guessed right. I'm George Stanhope of East Kansas Power. We continue to furnish electric power at a price that's affordable and less than wind or solar. Not many folks can afford to pay more for their heat and air conditioning. And we do it without the risk involved with nuclear reactors."

Jefferson from the bank added from his seat at a front table, "I second that. Our farmers and homeowners will suffer if energy costs cut into their ability to finance loans and mortgages. We try very hard to avoid foreclosures, but there's a limit."

Farmer Jones had another comment. "By Golly, I must say that life on the farm is a lot better than it used to be. We've got nice big tractors to do the hard work, thanks to you Jeff. And things like a big plasma TV in the living room. But you know, I can't do that on a quarter section any more. I have to put in a lot more hours on a lot bigger acreage. We've got a pretty nice standard of living, but we have to pay for it! Maybe we should cut back on some of the frills like big cars, and take better care of our land and the air. Wouldn't want to suggest you give up steak dinners every evening though!"

Jefferson replied, "You're right; we've worked hard in this country to use our natural resources to live a good life. I know what you are saying about the atmosphere. But it's those huge populations in the developing countries that are going to be the problem. Why should we make sacrifices when they are making the problem worse?"

Calvin objected mildly. "Who can blame them for wanting our standard of living. The problem is global; maybe we should set a good example to find the solutions."

Farmer Jim Anderson stood to get attention. Standing with thumbs under his overall suspenders, he complained politely, "You know, I never had a chance to go to college. These molecules and photo-whatcha-ma jigs are beyond me. I have to leave these things to the experts like our government officials and scientists. Now Professor, what's the bottom line— and what should we do about it?"

Calvin had a slight smile, then a grimace as he turned aside. "Now that's a hot potato! Let me say that there is no doubt in my mind of the validity of the scientific measurements. The radiation properties of the molecules in the atmosphere are also well known. On the other hand, the theoretical models are extremely complicated. For example, that greenhouse on the edge of town doesn't just sit and get warm. The plants would die if it gets too hot; we use fans and open windows to control the temperature. Our planet does this by transporting heat from the tropics to mid and high latitudes in atmospheric circulation, but there's no way to pump it back out into space. Now you may have noticed from the TV weather people that the New Englanders had a rough winter. And they show the jet stream coming from way up in Alaska down to Oklahoma and back up into New England. That's a pretty extreme mixing of tropical and arctic air. One might expect that sort of thing if global warming requires more atmospheric transport. The theoretical models will have to get that right!"

"Are you saying it is too hard to predict? Maybe we should just wait and see what happens."

"That would be a definitive way to go, but we might not like the answer. And it will be too late for us to correct the situation. We shouldn't procrastinate too long; the experiment is running! The data indicate a warming trend. And our industrial society has definitely tilted the earth's climate in that direction; the amount is somewhat uncertain. So what should we do? I've mentioned some of these, like planting trees and using alternative energy sources like solar, wind, and nuclear instead of fossil fuel. And if we don't do a thorough job of that, we better plan to adapt to different climate situations."

Harding was not satisfied. "That's not a very clear answer. We pay a lot of taxes to support our scientific laboratories and universities. All we get is maybe for an answer. Now our elected officials make definite decisions. They have my vote; I don't need to listen to these wishy-washy arguments."

Crause was annoyed, "Damn it! If they make the definite decision that blatantly ignores the scientific data we will all suffer."

Calvin gestured for attention, "I must apologize if you believe I have wasted your time; I warned you that I was not an expert. However, there are certain fundamental scientific matters that can be understood with a reasonable dedication of time and analysis. The subject is of sufficient importance that we should not do less. I hope that I have made some small contribution in that direction. I might add that there are some rather entertaining programs on public television that present some of this science in an understandable fashion—like Nova and Scientific American. I suggest that this kind of time is likely to yield a vote that is at least of equal importance as political speeches."

Calvin put his notes aside and commented, "I thank you for your attention. I've found your comments stimulating. And now, watch that thermometer!"

Chairs were moved noisily from tables and "Got to get back to work" when the Mayor struggled to his feet again. "Before formal adjournment today, may I remind you that next week's guest will be Congressman Jenkins speaking on his plan for privatizing Medicare. See you then!" Then turning to Calvin, he extended a limp paw to say, "Thanks for the talk. As you see, we have some strong differences of opinion about the economy and the environment."

Calvin wryly nodded agreement, then paused as the elderly farmer Anderson approached "Say now! You don't really expect an old dog like me to learn anything new about science, do you? I watch the TV weather report of course, but never really learn anything. They just read the temperature predictions off the map for me; never explain why the last predictions were off by 10 degrees. But I'll try to find that PBS channel. Mabel has me trained to watch sitcoms every evening though. Don't know if she will let me switch!"

"Jim, all you folks have the intelligence to pick up some basic scientific facts. For example, I'm sure you've noticed that the wind blows counterclockwise about a low-pressure area; and you know how to predict the weather from watching the wind. And you know better than to accept the pretty TV lady's remark that the winds will blow the clouds into the low center. You've made a straightforward analysis—and that can be applied to political comments too!"

Calvin put on his coat and walked to the door with the farmer. As Jim climbed laboriously into the ancient Ford pickup, he watched as Cal approached the Corvette. "Yes, I know you think this is the result of my inflated professor's salary. Not so! I blew some of my folks hard earned investments on this. Before Mom died, I drove a 10-year old Toyota!"

Retirement On The Farm

Calvin drove the Corvette smoothly along the Farm-to-Market road eastward out of town past the open pastures with dairy cows and the cemetery where his parents were buried. Then there was the long downhill glide towards Raccoon Creek. Approaching the two-lane bridge across the creek, he took the narrow gravel road to the north. A bright green field of spring wheat lay to his left; to the right, the black oaks and cottonwoods were widely separated in the grassy border of Raccoon Creek. Calvin had leased the cultivated fields to an industrious neighbor Swede; he had elected to let the riparian area return to a natural state.

The short lane at the mailbox with Carpenter attached sign led him to the right past the farmhouse to the implement shed attached to the small barn. There he parked the Corvette alongside the three-quarter ton flatbed truck formerly used to haul sacks of feed to the pigs and chickens living closer to the stream. Calvin felt some slight guilt in flaunting the Corvette in town. But somehow the arrival in the old truck to give a lecture on Climate Change might not generate much respect.

So he had continued his duty to repay a college education supported by his parents. But the results were less than satisfactory. The topic was too complex and controversial. Some in the audience were loathe to think about departing from the status quo of a generous standard of living in order to heed the scientific call to care for the planet. "It's too hard—don't want to think about it!" The majority couldn't care less; let the government take care of the problem!

Calvin checked that the Corvette's interior was immaculate, locked the doors, and trudged wearily to the house. Hopefully, there would be enough Scotch in the cabinet for a leisurely drink to go with some Ritz crackers and Swiss cheese. But before settling in with the snack, a quick stroll through the old dining room to check e-mail. He lived alone; the non-functional dining furniture had been moved to storage on the second floor, and the space devoted exclusively to a fairly elaborate computer facility with custom designed Dell, Hewlett Packard printer and scanner, and connections to Wireless Internet. A scientist was expected to have such a set-up; Calvin made little use of it though. Tropical fish were swimming gracefully on the monitor. Cal continued to the spacious living room with its plasma TV, Sony DVD and CD, Denon amplifier, Bose speakers, and attendant La-Z-Boy. The Kansas City classical music station was airing Mozart. Calvin expected nothing important on TV until the PBS News Hour.

The hi-tech trappings of the farmhouse were, like the Corvette, the product of inheritance from his farm parents. There were no other children. Calvin had lived frugally at the college for many years. There had been small but careful retirement savings in TIAA. Now the pension was adequate for his living expenses and life style. Of course, there was little opportunity for profligate living in Southeast Kansas. Travel tours were of slight interest; he had no desire to travel alone with others killing time in their golden years. He had thought about Nature tours to Costa Rica or Midway, but hadn't made the plunge. Perhaps a road trip with the 'Vette through the Rockies visiting bed and breakfasts might be scheduled for the hot summer months.

Calvin had just two years of marital bliss with a beautiful loving mate in his initial years at University. Mary had succumbed to cervical cancer. His devotion to her memory precluded any other romantic endeavors. He was quite aware that his reclusive behavior left little possibility of what might have been a normal academic life. He was generally known on campus as an old bird of a hermit and the reputation had followed him back to the farm as well. He had early on retreated to a world of books. Today, behind the dining room was an extensive library of read and reread histories and novels. The adjacent bedroom furnished a comfortable retreat with a book.

After Jim Lehrer and the News Hour, Cal ran a Lean Cuisine through the microwave and finished it off with a conservative helping of Haagen Dazs vanilla. Then with a cup of green tea, he retired to the bedroom to reread his copy of Ellis Peters' *A Rare Benedictine*. After a dozen pages, he indulged in a few minutes reverie of departed love, then drifted into deep sleep. Somewhat after midnight, he experienced the usual need for a bladder run. But then sleep was light for a time. This was interrupted by some faint outside sounds and finally punctuated by the sharp roar of accelerating motorcycles. This was unusual in this neck of the woods, but with renewed quiet he returned to sleep.

With the early dawn, Cal walked stiffly to the kitchen to make coffee. He was arrested by the unpleasant odor of hydrogen sulfide and searched for a cause. Finally, as he stepped out to the veranda, increased odor along with eggshells and yellow splotches around the door made the explanation clear. Rotten eggs! Disgusting! Vandals! But why? His lifestyles had never been of concern to anyone before.

Later, after hosing off the debris, he communicated his displeasure to Sheriff Brown. The reply, "That is unusual. We've had no similar problems for years. But I have a theory. You riled a few people with your talk at the Community Center. With that kind of attitude, you'll just have to live with it. Can't help you. Wouldn't want to spoil my support from folks here who like things the way they are." Cal was annoyed. Clearly the sheriff couldn't be depended to discourage the vandals. What to do?

At lunchtime he returned from his errand at the county dog pound to stop at Ben's Café for soup and sandwich. Ben drifted by to comment on his view of Cal's truck with a large black beast tethered on the flatbed behind the cab. Cal explained, "Yes, got the Newfy-mix at the pound over at Masonville this morning. He ought to discourage my nighttime vandals, don't you think!"

Then before returning to the farm, he had a kinky little idea inspired by C.S. Lewis' *Out of the Silent Planet*, and stopped at the hardware store for stencils and paint. Out by the barn at home he painted "DO NOT FEED THE HROSSA!" These were Lewis' large, black, water-loving creatures—a perfect description of the Newfy-mix. Hross was singular, but Hrossa had a better ring to it. Of course Calvin might have adopted another of these shaggy dogs. But then he would have a pair of dogs instead of a friend. Perhaps his nighttime marauders might pause to wonder: "Maybe this ain't just an ordinary Newfy-mix!"

The county dog pound had made Cal a present of 40 pounds of Dog Chow in response to his gift of $200. After the beast had indulged in lunch, he made the proper conclusion that Cal was Food God and remained in the close vicinity without restraint. Cal decided to christen him for the role abandoned by the elected official. "Sheriff, come sit!"

It's Later Than You Think!

Cal had discovered a book title on amazon.com that was intriguing. *The Professor* struck a familiar chord, of course. Not a biography though, but fiction. Nevertheless, he thought to order a copy, then noted its availability in e-book. And there were ads for e-book readers. Perhaps he should indulge in this hi-tech adventure in reading. But best have a hard copy on the shelf anyway.

A few evenings later he was inspired to give some further thought to the fictional character of Chris Baldwin, Environmental Liaison Assistant to Professor Mandryka of Golden West University. Chris' duty in the novel was to serve as communicator between the Professor's laboratory and various public groups, especially school age children. Perhaps it would be a neat idea for Chris to take over the duty of transmitting a fundamental awareness of the problem of global warming.

Calvin's experience with members of the Chamber of Commerce indicated a general lack of interest or awareness of the scientific facts of the greenhouse effect or climate change. His talk had identified the problems but accomplished little except to stir up some animosity. While he felt he had the information and the teaching skills to communicate, it seemed obvious that few of this group were willing to listen. There was also an uncomfortable feeling that perhaps his personal appearance did little to support a reputation of an academic intellectual. In any case, his words of wisdom were being ignored. Now suppose that Environmental Liaison Chris Baldwin were to make a fresh start. He could become a champion of scientific truth and vanquish the economic bigots and religious zealots.

There was once a fictional masked swordsman who called himself Zorro (the Fox) to right the wrongs imposed by the villains of early California. Calvin began to visualize himself as the Zorro of the world of climate change. If Chris Baldwin had such a website, he might find a more extensive and welcome audience. Cal became inspired to make use of this fictional character to ghost write some of the basic science relations of the atmosphere. "Zorro will ride again with his sword of Microsoft Word to expose those who have violated their privilege of enjoyment of Earth's environment!" This Zorro was unlikely to be unmasked by the very few familiar with the novel *The Professor—Chronicles of a University Life.* Perhaps an approach from such a fictional communicator with apparent science credentials would provoke some dialogue on the fundamentals. Calvin had the fundamentals; they just didn't seem very impressive coming from a scruffy senior citizen. He would sleep on the problem and give it priority attention first thing in the morning.

After standard attention to the morning bacon and eggs, Calvin poured the second cup of coffee and began to concentrate on his project to ghost write scientific information on global warming. The target audience was that large majority who felt they could ignore the problem; let the government take care of it. Unfortunately their logic in selecting their Congressional Representatives did not include scientific expertise. Considerations of religion, jobs, and the price of wheat were dominant. Thoughts about science were absent from kindergarten to Capitol Hill and the White House.

So, Zorro would have to be provocative and entertaining. A classroom lecture wouldn't work. But by the same token, he wouldn't have to describe every single detail of atmospheric science in the messages. Just get that large segment of the public to pay attention and think about the fundamentals. So, after obtaining the website attributed to Chris Baldwin, Calvin published an introduction.

> "Chris Baldwin here. I'm the Environmental Liaison Assistant to Professor Mandryka of Golden West University, here to discuss the subjects of global warming and climate change. You will have found that many individuals are eager to express their opinions on these subjects. One needs to examine their agenda. The CEO of a coal or oil company cannot be expected to suggest that you should curb your appetite for generation of energy from fossil fuels because it will produce disastrous effects to the earth's climate. Nor will an Audubon Society officer admit that our violation of the laws of Nature will have trivial effects on the habitats and lives of our feathered and furry friends. And the pronouncements of a politician require examination of the contributions of his constituents.

> Let me explain my agenda. My comments will be based firmly on scientific principles and evidence. Professor Mandryka expects me to give a simple explanation to my high school students of the somewhat complex atmospheric phenomena studied in his laboratory. Prof has also perceived a societal need for enlightenment with regard to an analytic approach to environmental problems. It has become my mission to remind my young students and their responsible elders of certain elements of

basic atmospheric science. Now these will furnish you the basic tools to take this analytic approach to some personal and political decisions about our planet. The level of my presentations should be suitable for those with a basic introduction to physical science. I apologize to those whose intelligence is insulted. However, I challenge you to apply this information to make a careful analysis of today's problems of the atmospheric environment.

This will be a series of communications. They will be archived at Golden West University for you to review at your leisure. In keeping with a need to awaken readers to understand and protect our environment, I have titled these messages "IT'S LATER THAN YOU THINK!".

"Finally, you cannot avoid the consequences of ignorance; you are part of the problem. Fail in your responsibility and you jeopardize the well-being of your children and grandchildren, perhaps even your golden years!"

Yours for analytic thought,

Chris.

Chapter Two

KATHY MARTIN, NEIGHBOR AND FRIEND

Neighborly Visit

Calvin's morning schedule usually began a few minutes before sunrise. Since an evening of serious reading usually led to a lack of concentration around 9 PM, he found that 8 or 9 hours of heavy sleep left his body in a state of wakeful discomfort. The increased light of a new day prompted arousal to start the morning coffee brew. The initial cup of coffee was accompanied by a study of the dawn sky conditions and a plan of the day's activities. After the second cup, he was sufficiently alert to attend to toilette and get dressed. His breakfast ritual of a bowl of cereal followed by a fried egg and toast required no thought. He usually followed this with an hour or so stroll down along the creek. This was sometimes delayed to permit evaporation of overnight dew; less frequently by a persistent warm-front fog or shower.

The dedication to a regular regime of exercise was rather erratic however. Having sat around in his college office or the Jamestown apartment, he had never been inspired to worry much about his body. But the good genes from his parents needed a bit of help. The college doc recommended walking to clear the arteries. Or swimming, but the Raccoon didn't look very inviting. "No jogging though, Prof—your knees aren't strong enough. You'll jar some of the cartilage loose." Calvin couldn't get excited about just walking around the block or through the shopping mall though. But a stroll along the Raccoon was a lot more interesting as well as constructive. And now he had a good buddy for company.

The morning schedule was altered slightly with the presence of his new houseguest. Sheriff expected attention at an early stage for a trip behind the barn for a deposit and a purposeful mark of a nearby cotton-wood. This was followed promptly with a return to the house for the expected bowl of Iams. After this, he made a more leisurely study of the perimeter of the house and barn area to detect any overnight visits by other four-footed creatures—usually raccoons. This accomplished, Sheriff returned to the porch to await duty to his master, the Food God.

Calvin found the company and attention of his new friend made an order of magnitude increase in the enjoyment of his morning walk. Sheriff inspected every bush and tuft of grass for information relating to any nighttime activity of other creatures. A raccoon's travel to and from the creek was of special interest. However, Sheriff's genes lacked the trail-kill urge; his meandering never took him out of sight of Cal.

It was a cool morning after an early evening shower that prompted Calvin to hike purposefully upstream along the Raccoon to study evidence of the response to rainfall. There was a slight increase in riffle activity, but the dark stagnant pools were unchanged. There were a few suspicious swirls indicating underwater activity. He wondered if perhaps a visit with cane pole and soft-shell crab would tempt action by a big channel cat.

And the water of the Raccoon was encouraging the grasses and bushes to a lush growth. Calvin suspected that there must be lots of bugs and worms living here; should be good pickings for birds. In fact, he could hear a robin sounding off in the distance—singing his little heart out for his lady. Sheriff stopped abruptly; there was a different melody coming from the clump of small trees just ahead. It was a beautiful song, but not familiar to Calvin. Must be something like a hermit thrush; should be a good place to nest. But the song ceased as Sheriff approached. Cal stood quietly but the bird was silent. There was a flash of yellow in the low bushes though. A little fellow was harvesting small bugs off the under side of the leaves. Well! Must find a bird book someplace and buy a pair of binoculars. Life would be just a little more interesting if he got acquainted with the locals!

Cal had walked most of a mile upstream when he decided to depart to the nearby road for a brisk hike home before being overtaken by mid-morning heat. He was slightly disturbed to find that his route would take him through the barnyard and garden of his neighbor's property. This was a bit embarrassing since he had neglected to check in with this nearest neighbor on his return weeks ago. On the other hand, it would have been customary for such residents to make a welcoming visit to a new arrival; he had a vague recollection of hearing of some health problems that might have interfered with such a visit.

Kathy Martin was sitting in the porch swing with her first cup of morning coffee. A not very large black cat hopped agilely up to join her, greeting her with a touch of white nose to her ear, then nudging the cup aside to curl up on her lap. The robin in the cottonwood by the house was responding to the dawn—his song a musical challenge to any red-breasted intruder with a similar territorial melody. It promised to be another damn fine sunny spring day!

Nightcap, the cat, suddenly lifted an alert head, then with arched back and erect tail indicated his concern about the distant activity down by the stream. Yes, the morning peace was broken. To the cat, the monstrous black beast with a rolling gait was cause for alarm and retreat. The dog was followed by a tall man, dressed in traditional Kansas garb—straw hat, bib overalls, and denim work shirt. As Nightcap disappeared into the kitchen, presumably to safe haven behind the refrigerator, his mistress felt some slight annoyance with the disruption. The morning was no longer calm—still the visitor looked harmless, probably a nearby farmer looking for a lost cow. Best be polite and friendly.

As Calvin approached the house, he noted the figure emerging to stand near the edge of the porch. He was startled to observe a significant departure from the expected appearance of an elderly farmwoman. The woman standing with cup in hand was tall, erect, and alert. She wore a dark jump suit that failed to obscure the female figure with full hips and upper body, accentuated by a token belted waist. Her head with straw-colored hair, poised on a slim neck, was highlighted by the morning sun. She had obviously noted his appearance and observed his approach.

Calvin stopped abruptly at this departure from the expected sight of a typical Kansas farmwoman. Then, in part to disguise his surprise, he called to Sheriff and secured him with a leash before continuing.

"Hello. I'm Cal Carpenter, your neighbor just south of here. I was out for a morning stroll along the stream and thought to stop briefly to say hello. I hope I'm not in unwelcome trespass."

Ah, yes! The retired college teacher down the road, but his appearance suggested that he had never left the farm. Must be neighborly though. "Oh, no. Certainly not. I'm Kathy. Come on in," was the prompt answer.

Calvin continued, "And this is Sheriff, my new companion. He's friendly, almost too much so. I'll keep him on leash."

Kathy replied, "Not a problem. I have a cat; he has wisely made himself scarce just now. Say this is a big guy!" This comment as she patted him cautiously on the head and caressed his ears. Sheriff responded with a casual tail wag and friendly stare from large brown eyes. Kathy continued, "Oh, say—would you like coffee? I have a fresh pot. We could sit and chat for a few minutes."

"Yes. I'd like that."

"Have a seat there. I'll be just a minute. Black? Cream or sugar?"

Calvin paused before sitting. "Cream and sugar if convenient, but not critical. I'd be grateful with black."

Calvin was puzzled. This lady was a mature 50 or so and obviously in good health. His vision of a retired farm couple was considerably inaccurate, and husband seemed to be absent. His recollection of the brief weeks in attendance on his dying mother was filled with visitors bringing casseroles and desserts. Those consoling visitors were elderly couples exhibiting signs of lives of hard work in the Kansas sun. "As you are perhaps aware, I've moved in to the family house after my mother passed on. I've been rearranging furniture and personal possessions of course. That has kept me rather busy, but I must confess to being remiss about making my presence known. Hope you will forgive me."

Kathy returned with the hot cup of coffee. Perhaps the first impression was misleading. Calvin may have adopted the casual Kansas garb, but his manners seemed rather formal. The ponytail and mustache suggested an aggressive personality, but his attitude was apologetic and conservative. Dress him in formal attire and he could be the college president—an attractive gentleman. But he was a scientist. Now she could picture him standing in open collar and white shirt sleeves directing students at a big telescope or maybe a cyclotron, something like that.

She began, "Of course, I understand. Now, it's Cal for Calvin, right? I remember reading of your talk at the meeting in town. And you're retired from teaching science at Jamestown?"

"Yes, and you seem to have me at a disadvantage. I believe we have not met previously—I hope I'm not mistaken."

"That is unfortunately correct. I lived here with my husband Albert for about 10 years; he died two years ago. We knew your mother well. Sorry we didn't meet you at the time of her passing; Albert was having an especially uncomfortable spell just then."

"Ah. That explains my failure at recognition. I imagined you to be one of the many neighborhood couples that I met at that time. I'm sorry about your husband though; I wasn't aware of his illness."

"Thank you. I understand that there hasn't been an occasion for us to meet. You see, Albert and I lived in Chicago where he was manager of an insurance agency. When he was diagnosed with cancer, we returned here to the family farm. Albert wanted to enjoy the pleasant time of his early years. Continuing to live in Chicago under those conditions would not have been good."

Kathy continued, "I've stayed on after his death. I'm really a part time resident. I teach music at University in midweek, then return for long weekends. I'm glad you came just now—might have missed you."

Cal had listened attentively to this brief history. He had also taken the opportunity to make a careful study of the lady. Her hair was nicely styled, short and tousled, a light brown with silvered ends. Eyes with light brown iris, medium large with very slight wrinkles fading away to the outside. Eyelashes were dark. Eyebrows were also dark; they angled slightly up towards outside, then curved abruptly down near the outer eyes. The resultant expression was alert with a hint of secret amusement.

Nose was modest, very slightly upturned. Mouth was generous with full lips. Chin prominent, but not aggressive. A slight sag in flesh at jaw line produced a dimple line extending from the mouth. Face was oval with cheeks wide, slightly prominent. Ears were small and neat with tops just in line below eyes and small lobes even with space between lips and nose. Calvin found he had an objective view that Kathy must make a pleasantly exotic figure on campus.

The object of this pleasant examination paused in her manipulations of cream and sugar to confront him with an accusing smile. Cal was initially embarrassed. Kathy had correctly assessed his admiring view and was delightfully amused. Best face the music. "Yes, you've caught me out. But you must realize that I expected to find a version of that classic portrait of an elderly farm couple. Please forgive me though. I should not have stared."

Kathy paused to reflect, "Ah, perhaps I should not have been so obvious about noticing. But then I must confess that I wanted you to know that I noticed! Now, shall we talk about the weather?"

"If you insist!"

"Really, I am pleased that you dropped by. I have friends in town of course, but neighbors are important. It would have been awkward for me after all this time to just drive up to your house and announce that I lived next door. Now I know about your lecture in town, but haven't noticed much activity about your place. What does a retired college professor do? I suppose you're writing your memoirs. Oh sorry, corny joke."

"Well, you know I don't have a lot to say on that subject. I guess I'm just wondering what to do when I grow up. Of course it has taken some time to just organize the house for a bachelor existence. My mother's furniture didn't fit somehow. And I haven't socialized much. I don't have much to say about the price of hogs and there's not a lot of interest in science."

"I see your problem. But the folks here are basically friendly, just takes time."

"I've become a creature of habit I guess. I'm not fundamentally antisocial, just act that way."

"Well, have to put a stop to that. People will gossip if they think we are feuding! Let's leave it that if you want to borrow a cup of sugar, just ask!"

They continued to sip their coffee in relaxed enjoyment of the springtime view. The conversation had clearly established that they were to accept the comfortable role of near neighbors. Calvin had no problem with that. Kathy was friendly and interesting, and pleasant to look upon. This provoked a pause to assess his personal appearance. In his academic career, Calvin had given little attention to becoming the scholarly image of a conservative member of a college faculty. Rather, his somewhat maverick character favored the overly casual dress and scruffy hair of his undergraduate days. Now it didn't seem quite proper that this lovely person should have to put up with such a scarecrow for a neighbor! Ah, well! His dedication to the memory of his mate of long ago precluded thoughts of any other close personal relationship and certainly not romance!

But, best to preserve a friendly atmosphere, "Your place looks neat. Lawn is happy with last week's showers. Lilacs are healthy; looks like you will have sunflowers later."

"Thank you. Of course we sold all our cultivated land to Murphys. I only have this little area bordering the stream. I don't spend time on it though. Jimmy Cantwell comes out from town once a week to mow the lawn and whatever. I once tried a flowerbed out near the road but it was pretty shabby. I asked for advice from Nancy and Jim Jameson who rent the red house at the farm a mile over. He works at the feed store. They told me it was hopeless. The big corporation that owns much of the farmland has used chemicals to control weeds; the vapors travel with the wind. My flowers qualified as weeds!"

Calvin nodded, "But you have a nice view of the woods along the stream. Your outbuildings have some history I expect, but probably don't have much use now."

"Yes. That old shed has odds and ends of hand tools and an old buggy—can't bear to part with it! We had a fair size barn but it fell down after we sold the hay. That was all that held it up. You can see that I am not much into farm activity—this house is my life. Oh, would you like to see—I'm rather proud of it!"

"Oh, of course. I imagine it's a lot different from mine!"

"This doorway is entrance to the kitchen, obviously. It's functional—lots of space. Breakfast is easy—sometimes get more creative with dinner. Now this is the music room, my pride and joy. That grand piano is more valuable than the entire house. Posters are favorite events from Chicago. Paintings are prints of course. The library there is a collection of musical scores and histories—references for my class at University. The messy desk is reserved for class preparations."

"Ah, I have one of those too—but don't have class as an excuse anymore!"

"Bed and bath in the rear. A spare bedroom upstairs—seldom used."

Kathy moved to return to the porch; after a significant pause she explained, "My life has been a day to day maintenance since Albert died. No long range plans. I have a few local children in for piano lessons on my day free from University. Very little natural talent, but it helps with groceries. I try to maintain some proficiency for occasions when I'm called on for support of recitals or other events."

"Yes, it's not good to abandon a life's talent."

"There now! Can't look back—the characters in white uniforms may be catching up! But now, I was joking about your memoirs. But I imagine you have some professional book or research paper in the works."

"Well, actually I must confess to simpler thoughts of a trip to some exotic place—just lying on the beach. But can't face the idea of becoming part of some tour group. Still I do feel some professional responsibility to transmit a scientific outlook to my neighbors. But my talk at the Chamber of Commerce wasn't that great a success. I'll have to give some further thought on how to go about it."

Calvin stood as he drained his cup. "I'm most grateful for your hospitality—morning coffee and conversation. It was good to meet with you; I'll try to be a good neighbor. Hope I haven't been too much of a disruption to your morning schedule. Best be moving along—see if I can't do something constructive back at the house, like wash the breakfast dishes."

"I'm happy that you stopped by. I'm here on most weekends, probably see you occasionally. Stop by for coffee anytime."

Kathy was pleasantly confused after Calvin's departure. Future weekends might be interesting times. Her associates at University were academic professionals; always wrapped up in their specialties. Social life there was hectic, but not exciting. Escape to the house along the Raccoon was always quiet and relaxing. The days here were full of memories of happy times, then of dedicated care for Albert's suffering. The marriage had been good. Time now to continue with her profession.

Still, the meeting with Cal had been stimulating. His professional background was obvious, in spite of casual garb of a farmer. Conversation was easy; he was not at all stuffy. There was the potential for some interesting conversations on daily problems—and maybe some philosophical ideas. Oh, yes! He would be back! Perhaps she could arrange to look a little more attractive. Now where did that thought come from—no reason to suggest any progression from being casual neighbors.

The following morning, after breakfast and a brief stroll with Sheriff to the stream and back, Cal positioned the laptop on the table by the kitchen window and arranged a comfortable seat. The view towards the shed with the truck and Corvette did little to inspire noble scientific thoughts. The prospects were sobering. This horse might be led to water but forcing it to drink would be problematic. Still, if some fundamentals of atmospheric science could be scattered about in an attractive manner, perhaps the reader could be enticed to think about the subject. Give it a try!

It's Later Than You Think!

This is the first in a series of messages to encourage you to make an analytic approach to matters of climate change. We will begin with some elementary ideas and vocabulary that can be developed into a relatively simple understanding of our physical surroundings.

Facts

Almost any textbook on the atmosphere can furnish more than you need to know in minutes. And many of these facts are so commonplace that we hardly notice. The air is transparent—unless we add particulates. Air is easy to move—yet it can exert enormous force. It is compressible—not so with liquids and solids. The air is a tenuous gas whose distance between molecules is roughly ten times the diameter of each molecule. It is mostly space! Molecules have energy of motion, colliding with each other in elastic collisions. The speed of any particular molecule depends on its last collision; a few are stationary, others move very fast. Temperature is a measure of their average energy of random motion. Pressure is a measure of the momentum transferred in collisions.

A molecule has mass. Gravitational attraction by the earth gives it weight. Atmospheric pressure is a measure of the weight of the air molecules above. And since air is compressible, density varies as pressure. So pressure and density are highest at sea level and decrease with height.

Add energy to air; the molecules move faster. Temperature increases. Collisions increase separation. Density decreases. The weight of hot air is less than cold air. Hot air rises. Not a miracle; density rules!

But if we move an isolated mass of air from high pressure at low altitude to low pressure at high altitude, it will push against the low-pressure air and increase in volume. It is doing work; this requires energy. That energy must come from the random energy of motion, and that is described by its temperature. So, the temperature decreases with altitude.

Let's complicate things. Add a substance like water molecules that can behave just like any other substance in the air, but take away some energy and it tends to stick together like a liquid. And then it can be fog, drizzle, or fluffy white cumulus. We see these sailing with the wind. But why do clouds float? Gotcha!

Or take away some more energy and we get a solid, like ice or snow. And because temperature is a measure of energy, a change in temperature is critical to the behavior of water in the environment.

How much energy? Your middle school kid did that experiment for you. Four ice cubes in a pan on the stove. Add heat and watch the clock. Note the time required to melt the ice, the time to bring the water to a boil, and the time required to boil the water away—to change it to water vapor. It took almost as long to melt the ice as to heat the water—and over five times as long to change to vapor. The times are a measure of the energy added. And if you reverse the process, you get it all back—conservation of energy!

Yours for analytic thought,

Chris

A Pleasant Habit

Calvin began to assemble notes on the subject matter for his ghostwriting project. He had no problem with the fundamentals of atmospheric science; those physical principles had not changed since his retirement. And there was considerable recent literature on global warming. Some was available in the scientific journals on the Internet. There were textbooks to be ordered from amazon.com. It might even be necessary to make a trip to visit the library at University. The days passed swiftly.

Sheriff found this neglect of long morning walks distressing. Finally, he moved ahead up the stream and sat waiting for Cal to follow. Cal recognized that his big hairy friend had appreciated the attention of the neighbor lady; and yes it had been pleasant. He followed; it was late in the week, Kathy would probably be home from University.

Sheriff heralded their approach at Kathy's house with "Woof! Woof!" and tail waving. Kathy emerged onto the porch with the greeting, "Sheriff, it's good to see you again. And would your friend be interested in a morning cup of fresh coffee?"

"Yes indeed. And hope you have a few minutes to visit." Sheriff avoided the leash and galloped up to the porch for a pat and ear massage.

"Cal, the coffee is always on for you. And Sheriff, it doesn't take much to get your attention, does it? One of these days, I must introduce you to my cat. I think he is getting curious about such a big black furry thing."

Conversation began with the weather and proceeded with an up-to-date on the recent activities. "I've dispensed with Beethoven and Brahms and let the movie, 'Amadeus' account for Mozart. My class can't wait to get the full treatment of the Beatles; it's a new definition of the classics. Now tell me the news on global warming."

"Well, my friends in Alaska tell me that the melting of the permafrost is getting to be a nuisance. They have to patch the roads frequently as the ice disappears underneath. And summer travel in the backcountry is hopeless with all the water and mud—and mosquitoes. Worse, the spruce and lodgepole forests are in bad shape because milder winters are encouraging the insects."

"Yes I see that on television. But, I suppose our Kansas neighbors think that's another world that doesn't concern them."

"Right! The Eskimos can use all that oil money to fix their roads!"

"I bring up the subject occasionally; that just ends the conversation. Are you doing any better?"

"Not yet, but I have a plan. Might tell you about it if it works. Don't hold your breath though!"

Calvin found the sunny view from the old rocker on the porch to be pleasant and relaxing. The grassy area under the cottonwoods was a pattern of sunny green and sharp shadows. Sheriff had found a position at the foot of the porch steps where he could watch for attention. Kathy was dressed in a flowery blouse with good curves and dark slacks exhibiting a bit of attractive leg. Most pleasant; might do this more often. Could get to be a habit!

And Kathy was giving him more than a little casual attention. She seemed to find Cal a subject of some real interest. Calvin became increasingly flattered. But he was puzzled. Surely he couldn't be physically attractive. Not that it mattered—there was no place in his retirement for romance.

"Cal, I'm glad that you and Sheriff are just down the road. I share a small apartment at University with a woman who has a similar life style. Her husband was a professor of classics who passed on a few years ago. I stay with her while I'm teaching and have thought to give up this place and be there full time. But I really do love the quiet beauty of this area. And it's good to know that I'm not alone with that. Now I baked some doughnuts last evening just in case. Could I tempt you with a refill if I bring them out?"

Yes, could get to be a habit!

Calvin hoped that those who browsed the Internet had found the introductory facts about the atmosphere. They would now be prepared to understand the concept of global warming and its relationship to power generation from fossil fuels.

It's Later Than You Think!

Think of an expression that should arouse guilt and frighten your grandchildren. No, it's not "War of the Worlds" or "Frankenstein". It may be the greatest threat to the environment since humanoids evolved. We call it "Global Warming"!

Global Warming

Now expressions like global warming and climate change have been bandied about for some time just as if everyone knew what they meant. People in the scientific community had a fairly precise understanding, but they generally don't communicate with the general public. Further information might be too taxing for an ordinary citizen. Reaction—too hard, don't want to think about it. Besides, the politicians are deciding environmental policies. But that means that such decisions are being based on ignorance and neglect.

And you, my intelligent reader, say you have been exposed to more than you want to know about the subject. So let's make some elementary checks on a few fundamentals. Global warming—what does that mean? Everybody's measurement of temperature increases? Scientists say that the global averaged temperature has increased 0.6 Celsius (or Centigrade) degrees during the last century. Averaged? How? And the thermometer outside your window reads 77 degrees Fahrenheit. What is that on the Celsius scale? And what is the change in Fahrenheit for 0.6 degrees Celsius? Now our public schools are not all that great, but surely you have enough science and arithmetic to answer those questions. Or your Congressional Representative can give you the answer. No? Let's go back to square one.

What is temperature? It's the number on the scale of
the thermometer outside your window! Well, it has something to do with the atmosphere. Right? We need to have
some model of the atmosphere. Now it is composed of
molecules of nitrogen and oxygen. At this point we don't
need to worry about their different structures. The air is a
tenuous gas whose distance between molecules is about
ten times the diameter of each molecule. It is mostly
space! Just imagine the molecules as identical ping-pong
balls that are having elastic collisions with each other.
Pressure is a measure of the momentum transferred in
those collisions.

Now you must know about energy; that's what your
two-year old has too much of! And molecules have
energy of motion, colliding with each other and any con-
fining walls. Temperature is a measure of that energy of
random motion. Now your thermometer might say 77
degrees Fahrenheit. Or it might say 25 degrees Celsius
for the same situation. Each scale is divided into equal
divisions between the freezing and the boiling point of
water—0 degrees and 100 degrees for Celsius, and 32
degrees and 212 degrees for Fahrenheit. (And that tells
you that a change of 1 degree Celsius is a change of 9/5
degree Fahrenheit.) But for the atmosphere, neither of
these is an absolute measure of energy; the atmospheric
molecules still have energy of motion when water
freezes.

Now we expect the pressure exerted by a gas to
decrease to zero when the gas molecules stop rattling
about. In the laboratory this happens when the Celsius
temperature reaches –273 degrees. So we define an
absolute or Kelvin temperature scale that is 0 degrees
when all random motion stops; the pressure and temperature are zero. If we use the divisions of the Celsius scale,
the freezing point of water now becomes 273 degrees
Kelvin. So now the energy of motion of atmospheric
molecules is proportional to the Kelvin temperature.

You may be forgiven if you've never heard of
Absolute or Kelvin temperature. (That's Kelvin, not
Calvin. Their callings were quite different.) But you
haven't been paying attention if you don't see that a
change in temperature, on any scale, means a change of
energy. And energy is important for children, puppy
dogs, geraniums, and practically everything on our
planet! Messing about with it could be hazardous to your
health!

Now, careful measurements for the past several hundred years have shown that the earth's average temperature has been gradually rising, and the rate is increasing.
And there are other lines of evidence that there is climate
change in response to human activities. We are reminded
that some fifty years ago, Roger Revelle and Hans Suess,
scientists at La Jolla, noted that we have been inadvertently engaged in an experiment in climate change. We
burn fossil fuels for power, resulting in an increase in
atmospheric carbon dioxide. This molecule is a strong
absorber of heat radiation from the earth, thus capable of
causing a general global warming. We are now becoming
concerned that continued burning of fossil fuels may
produce a disastrous climate change.

Yours for analytic thought,

Chris

Dinner Invitation

It was Thursday evening after microwaving a chicken thigh and green peas that Cal received the call from Kathy. "Cal, I hope I'm not interrupting your evening's entertainment. I found this number in the book under your mother's name. I suppose all your communications are done by e-mail these days, but I thought this needed the personal touch."

"Oh, sorry about that. And it is good to hear your voice. What's on your mind?"

"I bought a rack of lamb before leaving University with the thought that you might like to share it with me. I could work up a pretty good dinner about six tomorrow evening. Have to freeze it otherwise—probably serve it to Sheriff if you are too busy."

"Oh say! I'm definitely available for that. I can find an old soup bone for Sheriff. What can I bring—salad or something?"

"Sure. Maybe get one of those ready made things at the market—unless you're feeling very creative. And come by a little early. You can watch my kitchen expertise."

Calvin made a quick trip to the town market, planning to follow Kathy's advice on the salad. But then seeing some other fresh things from warmer climes, probably Mexico, he added some radishes, cucumbers and tomatoes to his purchase. His preparations would not be beautiful, but she would know that he had tried. And probably, to indicate his appreciation of this welcome gesture of a dinner, it might be wise to persuade barber Joe to give a neat trim to the sideburns and ponytail. Then, at home he gave some attention to his meager stock of wines and was happy to find a bottle of Montana Red from New Zealand South Island. It would go well with the lamb; that might be from the South Island as well.

Time passed slowly during the afternoon. His mind strayed from the global warming project; this dinner was an unusual and welcome event. Finally at twenty past five he retrieved some informal dinner clothes from his college wardrobe and polished his black Florsheims. There was a concern to look properly presentable but not too extreme; no need to put Kathy into shock. The Corvette would be a final touch, but Sheriff wouldn't fit; the truck would have to do.

Kathy was warned by the noisy truck and welcomed them at the kitchen door. She smiled at his unusual appearance and queried, "There must be a dinner party somewhere. Right?"

Calvin deposited the salad makings in the kitchen. Then extending the wine for approval, he found Kathy standing expectantly in his personal space. A greeting kiss on the cheek was in order; and the subtle odor of an attractive perfume caused an additional moment of hesitation. "Ah, now I should open this and let the wine do its thing. Then we can sample it while you keep a check on the food."

Sheriff had been reluctant to enter the house. He sat attentively outside the door awaiting greeting from Kathy. "Say, big guy. Smells good doesn't it. I hope Cal has brought that soup bone." Then with arm extended and hand signal, "But stay Sheriff. We'll join you on the porch with our wine in a few minutes."

Kathy was wearing a cocktail style apron over white blouse and dark slacks. A gesture to formality was apparent in the small turquoise earrings and neat matching necklace. The dinner atmosphere had taken on a relatively formal, but relaxed feeling.

Conversation progressed casually from weather to university affairs to thoughts about global warming. Then a move to the dinner table and the salad with side of Neuman's Italian. The main course with lamb, red potatoes, and broccoli provoked close attention and admiration. Finally, a small dish of Ben and Jerry's Cherry Garcia appeared. Calvin remarked, "Kathy, I hope this is a sign of more wonderful dinners to come. Life is taking on a new dimension!"

"Oh, I get these inspirations occasionally. Not terribly often; you'll have to be patient. Now leave the dishes. Coffee is about ready; Bailey's Irish Cream OK? Then we can sit on the front porch and watch the sunset."

Calvin adjusted the Adirondack chairs to continue attention to Kathy, with a general view to the northwest. "You've arranged for a beautiful red sunset."

"Well, yes. Takes careful planning—and a little luck! Life is good sometimes."

"That's true. We have about all one could ask for. The very best of food and good company." This said with a gesture and smile to Kathy. Then, "And, can't beat the good old U.S. of A. It's a good country. Not only beautiful, but gives us tops in standard of living with shelter, clothes, and all sorts of toys to entertain us. It's no wonder that we are envied by others. There's immigration pressure of course. And then there are some who would just like to bring us down to their level."

"I know. I've enjoyed the best in culture, music especially. Can't do that if you worry about where the next meal is coming from. We do worry occasionally about having someone destroy our good life. But there are those who have to fight every day for what little they have."

"But when I see on TV how some of them live, I feel a little guilty. Actually it's an accident of birth. I suppose we could make a correction— send them money, share our food, give them our technology. But, I don't really want to give up this standard of living. And it's obvious that such an idea is not popular. So, we just thank our lucky stars and ignore the situation."

"I do try to keep our musical heritage alive. At least the next generation here will have it. Maybe it will filter out to the rest of the world, but can't imagine how. That's probably being egotistical; other places have their culture, like beautiful architecture. And things get stolen; music, literature, even buildings! And there are places that have next to nothing and have no hope of improvement. Still, it seems to be the intermediate culture that is dissatisfied and competitive."

"And our culture is pretty aggressive. But we seem to want more of the wrong things. Bigger cars, weapons, whatever. Our culture isn't improved; the world isn't any more peaceful or gentler. I think we are losing an opportunity that comes with being a wealthy nation. We should be a wise leader. If we are setting the wrong example, others will follow and the whole planet will suffer. I'm thankful for my science education; I can understand those things. I can try to improve the science outlook, but not hopeful for improving the big picture. Still, if no one does their bit, the situation will never change."

Calvin paused; his thoughts became more personal. "Of course, I keep thinking I have an excuse. Life isn't fair. So I just gave up on things beyond science after losing Mary. Oh, I suspect you knew I lost my wife to cervical cancer just a few years into our marriage?"

"Something like that—didn't have the details. And haven't wanted to pry."

"Not a secret of course. But haven't had much of a life beyond science. Even that is pretty average. I've just given up. But then I sometimes wonder if something will happen to change my outlook." Then staring at Kathy, he thought, "Maybe something has!" Glancing away to Sheriff, he gave a careful sigh, thinking, "But this gorgeous lady isn't about to find an old retired classroom teacher an attractive lover."

Calvin and Kathy sat silently then, watching the twilight glow, each with private thoughts.

Finally, "I can still hear Sheriff gnawing on that bone. But I expect he thinks he has been deserted—better head on home."

"There is some leftover lamb. I can add a few veggies to that and you will have lunch for tomorrow. I'll wrap it up for you."

Kathy maintained a cautious attitude. Before passing the food package to Cal, she made with a careful hug and an intimate pause. "Good night, Cal. Thank you for coming—and coffee will always be ready any morning."

Calvin experienced an unexpected lift in spirits. But in choking up, he could only mumble, "Good night."

Calvin was anxious to follow up his introduction of global warming with an explanation of the common expression, "The Greenhouse Effect". Best get into the nitty-gritty of the subject before his readers drifted away!

It's Later Than You Think!

Let's understand this talk of the Greenhouse Effect. Is this a house for little green men? Or is raising orchids about to destroy the planet? Time to get to the nitty-gritty. Then we can all start on the same page of this problem.

Greenhouse Effect I.

The Greenhouse Effect is said to be responsible for keeping our planet at the proper temperature. What is it? The name implies it is similar to the operation of our greenhouses that we operate to have a warm place to grow vegetables or flowers. You have seen these buildings, perhaps actually been inside. And since the greenhouse effect is so important for our climate, I'm sure you have investigated and know exactly how these operate. If not, I'm sure you have the best of intentions to correct that situation. I will help you.

These often have glass roofs that transmit visible sunlight to the plants and soil, but absorb their heat radiation. The outgoing heat radiation is trapped. The temperature increases. Now, you really don't believe you get heat radiation from black dirt, do you? Have you personally made that observation? Another science myth? Ah, but you can see the red light from the hot coals in the fireplace and you may have felt the heat radiation from a warm object, like the household iron used to press clothes. To observe the far infrared radiation you just need a more sensitive detector!

Actually, if the air is confined in the building, the warm air is trapped; there can be no convection to distribute the excess heat up into the atmosphere. The temperature may become too high for the plants. So we open some windows and turn on some exhaust fans to regulate

the temperature. And conditions inside the local greenhouse depend greatly on the outside weather; for example, the greenhouse temperature will not be so high if
there is a cold blustery wind outside.

The earth's atmosphere functions somewhat like the
glass in our greenhouse. It transmits most of the visible
sunlight and absorbs a relatively small fraction of the
sunlight at certain wavelengths in the infrared. On the
other hand, the outgoing radiation from the earth maxi-
mizes in the infrared, just where the atmosphere is
strongly absorbing. A large fraction of the incoming radiation arrives at the surface; only a small fraction of the
outgoing radiation escapes. The molecules that are effective in the absorption of the outgoing infrared radiation
are called greenhouse gases. Now our planet operates in
a more restricted fashion than the greenhouse a few
miles away. The radiation balance is supreme. The earth
is isolated. There is no other way to get rid of any excess
heat; no fans to blow the heat out into space! But we can
diddle with the radiation.

Now what is the difference between visible light from
the sun and the heat energy from the warm soil? Physics
books don't read like romance novels, and the lovers
don't tell you that light from the sun and heat radiation
are part of an electromagnetic process with a continuous
spectrum of wavelengths from ultraviolet to infrared and
radio waves. The radiation spectrum of a heated object is
related to the vibration of its charged particles. Those
vibrations vary in frequency as the energy of the particles. (Actually, the relationship between energy and radiation frequency is a very precise proportionality and
represents the birth of the quantum theory of matter.) The
energies of motion are continuous and the average
energy is described by a temperature. So, the frequency,
or the wavelength (or the color), will be continuous from
ultraviolet to infrared and there will be a maximum
intensity of radiation at some wavelength determined by
that temperature. The relationship is surprisingly simple;
the product of the wavelength at maximum intensity and

the absolute temperature is a constant for radiation from all heated objects.

For example, the sun with a surface temperature of 6000 degrees Kelvin emits light of all wavelengths, which our eye sees as white light. The maximum intensity is at a wavelength of about 0.5 microns (1 micron = 10^{-6} meters) in the green region of the spectrum. It follows that objects like the greenhouse soil at room temperature of 300 degrees Kelvin emit electromagnetic energy—heat energy—with maximum intensity at 10 microns in the infrared. (Check my arithmetic: 0.5 x 6000 = 3000 = 10 x 300!) The structure of glass is such that it transmits visible light at 0.5 microns, but absorbs infrared radiation at 10 microns.

Now, most physics and chemistry books will not bother you with the absorption spectra of the molecules in our atmosphere. But trust me! (Trusting a scientist can be no worse than with a politician!) It is this molecular absorption of heat radiation that forms an atmospheric blanket. Without this blanket, the balance of radiation would result in a colder earth, less favorable to life. The water molecules are effective infrared absorbers and the large amount of water vapor makes this the most effective greenhouse gas. Some other molecules have lower concentrations, but have high absorption probability and are important. Some of these are carbon dioxide, methane, ozone, chlorofluorocarbons, and nitrous oxide (CO_2, CH_4, O_3, CFC, N_2O).

The amount of each of the greenhouse gases is influenced by human activity. The total amount of water is too large to be changed appreciably. (There are some secondary effects relating to ice, liquid, or vapor that are important though.) Most important of the others is CO_2, which results from the burning of carbon containing substances like fossil fuels. This gas has increased by about one third since the beginning of the industrial age.

So technology solves all these little problems. You're going to amend a few laws of physics and radiate the extra energy back into space at some other part of the spectrum. You have a procedure in mind? I'm sure big government will ignore the science and subsidize you. Do it!

Yours for analytic thought,

Chris

Early Tornado

The polar front had retreated northward with the sun. Southerly breezes brought moist air from the Gulf. The high midday sun warmed the farm fields and generated cumulus clouds that grew high in the atmosphere. Showers grew more frequent in the late afternoon.

Calvin arrived for morning coffee an hour after sunrise. "Hope I'm not pushing your schedule too early. It's refreshing to walk early; gets hot quickly these days."

"Cal you're fine. I'm an early riser too; it's the best part of the day—shame to miss it. Afternoons are getting sultry—good time to read, or take a nap!"

"We're probably due for some cool Pacific air any day. Weather might start getting rough. Keep an eye on the sky if there are predictions of violent thunderstorms or tornados. Hate to think you might have to go to the basement, but be prepared."

"Now Mr. Rainmaker, no need to beat your drums. Farmers are busy making hay. Wouldn't want to get them upset!"

Calvin continued to stare at the growing cumulus as he hiked briskly with Sheriff towards home. The cloud bases seemed to be getting dark earlier.

The cool air mass edged slowly in from the west. At high altitude, with less resistance from the ground, it bulged ever so slightly over the warm moist air. The afternoon thunderstorms formed a jagged line that grew ominously into the evening. A violent weather alert was issued. Farmers moved machinery from under trees. Town folks gathered up the lawn furniture. The storms arrived locally with gusty winds and heavy rain late Saturday night. The cool dry air arrived early for a gorgeous Sunday morning. Calvin's first thought was to make a phone check with Kathy for any need to help with storm damage. "Kathy, how did you make out last evening? Hope you didn't have any damage."

"Good morning, Cal. Nothing serious here—a few tree limbs scattered about. Your vehicles' safe? Wouldn't want that Corvette to get dented!"

"No problems here. I'll be up in a bit to clean up your tree limbs— and coffee—doughnuts too?"

"You drive a hard bargain! See you!"

The telephone call arrived during the coffee break. "Kathy, this is Lois with the church group."

"I'm here Lois. You sound like bad news. What's happened?"

"Yes. We just had word that a twister hit that house on the edge of Jackson's farm where the Chavez family live. Sarah suffered a broken arm but the rest of the family is OK. But it demolished the house and scattered their things to hell and beyond."

"Oh my! That's rough. What can we do?"

"Well, they didn't have much, but now nothing. We thought to take a collection to buy the kids clothes and whatever. Jackson had medical insurance for them so that's not a problem. Pastor Swenson has scheduled a brief meeting Thursday night to get things rolling. I wondered if you would be back from University in time to help with the service like you do occasionally?"

"Of course Lois. I'll be there. Talk to you later about details."

With Kathy off the phone, Calvin remarked, "So—somebody got hit!"

"Yes. The immigrant family who work for Jackson. House was destroyed. Mrs. Chavez has broken arm. Gomez and kids OK but personal things are gone. Church will meet Thursday to take up a collection."

"Tornados don't care who they clobber do they! I'll try to remember to be there Thursday."

"Tie a string about a finger. See you there!"

Well, hopefully the last Zorro message about the greenhouse effect has exposed some lack of thought about these matters. Now we can move ahead to talk about serious problems. But that little storm the other night presents an opportunity to remind readers what the Weather Service has been trying to tell folks about their environment.

It's Later Than You Think!

So, tornado alley is getting active again. You would think the government would give some of FEMA's money to the Department of Energy so they could fix that. Oh, they've already done that? I've got news for you and your politicians. Mother Nature plays by a set of rules that can't be changed. We've got meteorologists— weather scientists—who understand those rules. When they analyze the atmospheric measurements, the conditions for violent weather become evident.

Tornado

The first danger signal comes with large amounts of water vapor. Now, there is a lot of energy stored there that can be released with condensation. So, when the ground is heated by the sun, the moist air is warmed and lifted to altitudes of lower pressure where it is cooled. When that pocket of air cools sufficiently, a cloud forms with the condensation. But the condensation energy release makes that air warmer than outside the cloud and the air continues to rise. The situation becomes more unstable if the air mass above is cold and dry; the cloud continues to be warmer than the surroundings. The cloud continues to grow so long as the supply of moisture is available. We have a thunderstorm with rain, possibly hail, and violent updrafts.

So when there is lots of water vapor, surface heating, and cold air above, we have an extreme situation. Then we may observe what might be called the ice-skater syndrome. When the skater moves extended arms inward to the body, the angular speed automatically increases. In our storm, the ascending column of air must be replaced by other surface (moist) air. Suppose the air at 1000 yards from the storm center is rotating slowly counter-clockwise at about 10 miles per hour. We say its angular momentum is the product of its linear momentum, mass x speed, times its distance from the center of rotation. If there is no external torque, like a few thousand feet above the ground, the angular momentum must remain constant. Moving that air from 1000 yards to 50 yards from the center will automatically increase its speed to 200 miles per hour. We have a tornado!

Now these tornado conditions are not quite that simple, and not exactly ideal for making measurements. Things can get pretty complex and violent and happen so quickly that observations and predictions are not feasible! It follows unfortunately, that the available information is seldom precise enough to predict exactly where and when the tornado will form. A tornado watch is broadcast when the general conditions are ripe; a tornado warning is called when the rotation is visible or detected by Doppler radar. The only personal defense is a sturdy safe room for refuge. Sorry about that!

Now I have explained that a large amount of water vapor is necessary for tornado formation. Test question: If with global warming the temperature of the air and the water in the Gulf of Mexico is increased, what will happen to the source of energy in tornado alley?

Yours for analytic thought,

Chris

The Voice

Calvin arrived at the church as Pastor Swenson was explaining the problem and the need. "~and the Chavez family is at the hospital with Sarah. She has a broken arm that is a problem. Now Chris Jackson has arranged for a double wide for them and furniture is no problem. But there is a critical need for clothing. If you have something appropriate that is in good shape, you can drop it off here later. Tonight we could use some cash to buy shoes for the kids and that sort of thing."

"Now to begin, I thought to express our feelings in song." The two-year old Bascom boy sitting beside Calvin took this opportunity to express his displeasure about close confinement and Pastor's speech was broken up. "~has consented to present the Lord's Prayer on our behalf. Then she will lead us in our song for the future, *'He's got the whole world in his hands'*; seems appropriate today. Then we will have the collection to help the Chavez family in their recovery."

This was followed by a few organ chords, and the congregation quieted. A tall woman, dressed in conservative dark suit stepped forward from the front pews, facing the altar and organ. There was a quiet pause after the final organ chord, then, "Our Father,~ The rich alto voice began quietly, then strengthened with "which art in heaven ." The singer stood facing slightly upward to the large stained glass window in the east. Her voice was strong and vibrant. The very slight delay in the echo made the Lord's Prayer seem to originate from the entire assembly.

Calvin stopped breathing. The proceedings were a surprise. Not so much to have a solo rendition, but the voice was very special; it filled the church, and sent a shiver down his spine. And there was something very personal in the singer's voice. He lifted bowed head to stare at the figure at the front of the congregation. "~and the glory forever". The organ added a gentle 'Amen'.

As the organ began a spirited interlude, the singer turned with a smile and motioned the audience to stand and begin the initial phrases, "He's got the whole world~

Calvin stood with mouth agape. The singer was Kathy! A few minutes later he was able to join in the happy expression of song. Then the collection plate made its way through the audience. Calvin deposited a small plain envelop with the thought, "That couple of hundred should put them in new shoes at least for the kids!"

He waited outside in the twilight as the congregation departed. Kathy was in the company of several church ladies as she stepped outside. Seeing Calvin, she moved toward him with, "Glad you came. I hope we did well for the Chavez family." And she was particularly gratified that Cal had been there for her brief performance.

"Kathy, it was a fine gesture. And you made it exceptional. Your voice is really special." Then with attempt to lighten the comment, he added, "Almost made a Christian out of me!"

Kathy confided, "That was what they wanted and needed tonight. Professional singers do that sort of thing you know." This said with a mischievous smile.

Then, glancing over her shoulder, "But, please excuse. I need to stay with the group for a time. See you tomorrow?" Cal nodded with a smile.

Calvin piloted the Corvette towards home marveling at the revelation of Kathy's beautiful voice. He had been quite satisfied to accept her piano expertise and knowledge of music history as sufficient to warrant her a faculty position at University. Now this voice suggested something more profound. He began to wonder at the other possibilities. Could there be recordings? There was something more familiar about that voice than just being from a new friend.

And there was something similar in his library of old vinyl records. There was the inspirational selection from "The Sound of Music". He found the track for 'Climb Every Mountain' and set it to play. The voice was lighter than Kathy's church performance, but startling similar. The album was a selection from a Chicago Theater production some years back. The list of performers included the Katherine Martin of yesteryear!

As Calvin approached the house of Kathy next morning, he found a cup of steaming coffee waiting. "Saw you coming. Sheriff, heads up!" A dog biscuit was lofted in his general direction; forefeet in the air, a loud 'glomp' snared the tidbit.

Calvin eyed Kathy in a new knowing fashion. She stared back with an amused, but too innocent expression. "Kathy, please forgive me for being so dense. Your beautiful voice was an embarrassing awakening for me. And you haven't been about to enlighten me, have you?"

"No special need. Knew there would be a time for you to know more about me. But was worried that you would forget to show up at the Chavez service! Ah, but all my songs are not that religious. I have a few lines from a seasonal one I'd like to do for you. Come stand by the piano."

Kathy seated herself at the keyboard and played a few chords. Then, "It's called 'Pieces of April'." A brief introduction and "April gave us Springtime." Calvin was again entranced by the voice. The light atmosphere of the lyrics was a pleasant change from religious fervor. Still it didn't matter; this was a personal performance, just for him. And he was captured by the happy eye contact accompanying the music.

Kathy was in her element. She was well aware of her talent. A friendly neighbor deserved her best. The music was for the present; thoughts of the future were pleasantly vague. "and it's a morning in May".

Calvin sensed that this was something of a personal communication, a sharing of things good. He applauded cautiously, "Kathy you are one talented lady—and I thank you for being so delightful!"

Calvin was uncertain how the last message on the Greenhouse Effect had been received. Perhaps it had been too abrupt an immersion into serious science. But, press on!

It's Later Than You Think!

I know, you've already read more about the greenhouse effect than you wanted to know. But hang in there—this won't hurt—much.

Greenhouse Effect II.

Unlike the greenhouse in your garden, the earth's temperature is determined entirely by radiation balance—no way to open the windows to space for ventilation. So, let's consult your 13-year old arithmetic genius to get some answers. He looks up the observed value of the solar constant (in metric units, of course). Then to get an average value for the incoming radiation, he divides by 2 to account for nighttime and divides by another 2 to account for earth being a sphere. Then he finds a simple formula in a standard physics textbook that permits a calculation of the heat radiation from an object at a certain average temperature on the Kelvin scale—that's outgoing radiation. Setting these equal on the back of an envelope, he finds the correct average earth temperature to be 280° K. (K for Kelvin—that's +7° Celsius, about 45° Fahrenheit.)

A bit of a disappointment. The answer is well below the observed average temperature for the earth's surface. And if he/she corrects for reflectivity of the sunlight, the answer is lower—below freezing point of water!

So, with a little suggestion from you, your young genius adds the greenhouse effect with absorption of infrared heat radiation by water vapor. With some finagling, a correction might be made to limit the infrared loss, permitting a higher earth temperature.

Careful now! Let's compare the earth situation with the heat radiation that you feel from the household iron across the room. But the earth is a surface like the entire wall covered with hot irons. That's a lot more radiation! And on the earth, much of that radiation is absorbed by a continuous blanket of water vapor in the atmosphere. This blanket is warmed and radiates infrared back to the earth. So, after things settle down, we can imagine that the layer between the sky and earth is filled with a swarm of infrared photons moving back and forth—they are trapped!

Your 13-year old now turns to use a laptop. A couple hours later he/she will probably tell you that the trapped infrared has increased the temperature well above that expected in the absence of the greenhouse effect of water vapor.

So H_2O is a greenhouse gas and it is helped by CO_2, CH_4, and others. Now there are some grown-up geniuses earning their keep by putting observed facts into super-computers, and now they tell us that if we keep adding CO_2 and CH_4 to the atmosphere, you will have an enhanced greenhouse effect; the earth is bound to get warmer.

Yours for analytic thought,

Chris

Chapter Three

HEAVY WEATHER

Monster Storm

With midday sun, the sea surface temperature at 10 degrees north latitude just off the African coast was increasing towards the upper 80s on the Fahrenheit scale. The number of energetic water molecules escaping from the surface was slightly in excess of those that returned—water vapor with its stored energy was increasing in the atmosphere. Warm columns of air were lifting the vapor to lower pressures where it was cooled to condensation in cumulous clouds. A few days later, the clouds were larger and more numerous. They were warmed by the condensation energy release and a surface low pressure appeared. Showers occurred more frequently with more energy release. A gentle counterclockwise circulation began and was noted on the weather maps. A light easterly wind aloft encouraged the clouds to move westward as they grew. The tropical storm intensified as more and more water vapor energy was added from the ocean surface.

About ten days later, the NOAA Hurricane Center upgraded the storm to category 1 hurricane as it entered the northern Caribbean. Then moving into the southern Gulf of Mexico, it found a surprisingly warm sea surface; the additional water vapor added to the storm's energy. That energy release lowered the central pressure and increased the eye wall winds. The residents and tourists on the Yucatan found themselves on the fringe of Hurricane Abe and were given a taste of Nature's fury.

The storm regained energy as it moved into the warm Gulf of Mexico. A category five hurricane was imminent. Residents of the Gulf Coast from Houston to Pensacola were warned. Landfall occurred just east of Houston with little energy dissipation of winds or storm surge. The offshore islands and marshes had disappeared over the years. Death and destruction made the news. Then, deprived of its ocean source of energy it was downgraded to a very large tropical storm; the system moved northward, carrying immense amounts of water vapor and its stored energy.

Calvin noted the simultaneous rapid approach from the west of a modified polar air mass. All hell was about to break loose. The National Weather Service was giving attention to the tropical storm and the potential for major flooding in the lowland areas west of the Mississippi and its tributaries. They had not yet focused on the approaching cold front and the related danger of violent weather. How bad was it going to be? There had been episodes in weeks past of numerous tornados occurring during several afternoons or nights in serial fashion. Destruction and loss of life had been considerable. That appeared to be the likely scenario.

But conditions appeared to be ripe for a 'monster storm'. There was a tremendous amount of energy available; the cold air aloft would add greatly to the instability; some preliminary surface heating was a likely trigger. Could this result in a single monster tornado or rebirth of the hurricane? It might be a one-shot explosion of energy release; there would be no way to replenish the water vapor energy as over the open ocean. But there would be an extremely large amount of sudden precipitation and resultant flooding. Cal was not aware of any such recorded event; but it appeared to be a possibility! And the Raccoon Creek region was likely to be near the western edge of the eye wall!

Calvin's farmhouse had not been constructed with a full basement; there was only a small area occupied by the furnace. However, there was a spacious storm cellar that had been excavated in the hillside a few feet behind the house. Calvin had always planned to seek refuge there if and when his weather-radio gave its standby warning. This situation suggested that he might be spending a significant amount of time underground. There was time for some preparation.

Kathy was at University for another day or so. But Cal had made no arrangement to communicate with her there. The best he could do was to leave a phone message to warn about heavy weather. "Kathy, I'm sure you are aware of that tropical storm and the likelihood of heavy rain. But the weather situation is setting up for potential tornado danger. Batten down the hatches and get your weather-radio on standby. And don't hesitate to move to a safe room or the basement. Call me if you need help with anything."

Cal had insurance on his house and personal belongings. Things like the computer and entertainment electronics were replaceable. But he would rather not see certain personal papers, pictures, books, and back-up files scattered about the countryside. Best apply some additional attention to the 'precautionary principle'. Selected items of the above were boxed and carted to the storm shelter. The inventory of reserve food and water was checked and supplemented. Sheriff's interest in these activities reminded him to include that extra sack of dog food. His camp stove and spare coffeepot with sack of beans were positioned for ready use. The folding camp cot and sleeping bag occupied the remaining space. He hoped this emergency living space would not be necessary, but the situation looked ominous.

By Wednesday midday, preparations for the storm had been completed. Cal constructed a sandwich of whole wheat bread, mayonnaise, lettuce, and chicken. He added a carton of yogurt, and a snack for Sheriff; then moved to the porch for lunch. Plans for dinner suggested a steak and veggies from the freezer; no need to let it spoil after a likely power outage. This thought lead to concern about the website preparation. Perhaps it would be good to prepare that before the storm. Hurricanes were on his mind.

It's Later Than You Think!

So, the greenhouse effect has trapped a little extra energy in the earth's atmosphere. And the midday sun is always near the zenith at the equator. So it gets a little warmer. Well, the natives can manage that; why should the rest of us worry? But storms that form in the tropics, hurricanes, typhoons, cyclones, transport tremendous amounts of energy and precipitation out of the tropics. Uh, Oh! Trouble!

Hurricanes

So, best take note when the clouds start to form over the eastern Atlantic. As that pocket of air moves westward across the ocean, it continues to accumulate water vapor. And the amounts increase if the sea surface temperature increases even slightly; the greenhouse effect is bound to do that! The air in the cloud is warmed by the condensation and the surface pressure is decreased below this column of warm air. As the surrounding air responds to this pressure difference, it must take into account the rotation of the earth, the Coriolis effect. Soon this air follows a path with the pressure difference force just balanced by the apparent force of the moving air over the rotating earth—in the northern hemisphere, counterclockwise about the low pressure center.

(Now I'm sure you are familiar with the Coriolis imaginary force invented by a mathematician to show the French artillery where to aim their guns. Ah, you didn't get beyond seventh grade arithmetic. Well, pretend you are a budding Einstein and do one of his 'thought experiments'. Imagine yourself to stand about one football field south of the North geographic pole. There is a pendulum bob hanging from the North Star and it is swinging toward you. It will hurt if it hits you. But you are safe; it will miss you; it appears to have been deflected to the right of its motion because you have moved eastward as the earth rotates. That is the Coriolis Effect.)

Now as this storm develops with the energy release from more and more water vapor, the central pressure continues to decrease and the rotating winds increase. The momentum of that air increases with the speed, of course. And that determines the force of a gust of wind on your house. Now I hope you also know that the kinetic energy (energy of motion) of the wind increases with the square of the velocity. (True for automobiles too!) So if a minimal hurricane forms with winds of 75 mph, at what wind speed is the energy doubled? Too hard?

Some elementary algebra: $S^2 = 2$ x $(75)^2$; take square root of both sides,

$S = 2^{1/2}$ x $75 = 105$! And tripled: 130! And quadrupled: 150! Or, get your 4th grader to show you with his hand calculator!

Now as the energy and precipitation of such a storm is dumped on Florida or the Gulf Coast, one may expect considerable immediate damage. And the continued transport and condensation of all this water vapor into higher latitudes inland can be expected to produce additional violent weather. FEMA will be busy spending your tax money, your insurance rates will increase, and you will spend a lot of time fixing things!

Yours for analytic thought,

Chris

The Storm Shelter

Thursday morning dawned with the orange sun adding heat to the thick haze. The wind was light; the humidity was stifling. The first rain band of the tropical storm arrived about midday. There was minimal precipitation; there was little cooling effect. The temperature remained in the 90s. A second rain band came in late afternoon. Towering cumulous clouds were building in the tropical air mass. The National Weather Service warned of severe thunderstorms and heavy rain. Evening dusk arrived early. Skies were dark with continuous cloud-to-cloud lightning. Showers were erratic with gusty winds. At 7 PM the tornado watch was issued. Television reception was hopeless. Calvin sat in the doorway of the farmhouse listening to news on battery-powered radio, Sheriff at his feet.

At 9 PM the weather-radio sounded off. Doppler radar had detected rotation. The tornado warning was in effect; time to seek immediate shelter. Cal breathed a long sigh, "Sheriff, it's time. I'm afraid we must retire to our storm shelter. Come!"

Arriving at the shelter, Calvin lit the kerosene lantern and prepared to settle in. There was a feeling of despair as he realized the house was vulnerable and unprotected. But he remained at the doorway until he heard the roar of the oncoming wind. Then, latching the heavy door securely, he sat in resigned fashion on an empty vegetable crate. Time stretched on; the roar of the wind continued unabated. There was no way to identify the snapping of tree limbs or the scream of collapsing walls and roof of the house. Frequent popping of eardrums signified the rapid large pressure changes. There were occasional heavy thuds of a large object impacting the heavy door of the storm shelter. Sheriff moved continually with nervous whines. "I know my friend; things are not nice out there. Come here, old buddy; we'll worry about it together."

The din continued. Calvin moved to the cot and slept fitfully with arm resting on Sheriff. Suddenly in the early morning hours it became completely quiet. Cal slept on. Sheriff awoke, eager to depart from the dimly lit interior of the storm shelter. A few impatient nudges aroused Cal who was initially confused with his surroundings. Then, realizing the storm had ended, he scrambled to open the heavy door. The cool damp air brought him fully awake. The predawn sky was solid overcast with low clouds moving rapidly to the south. Cal's heart sank as he realized the familiar profile of his house was no more. With a sob, he knelt beside Sheriff, "Damn, it's not fair." Then with a long sigh, "but it's exactly as we expected; we'll make out somehow won't we old fellow!"

Rescue

Suddenly, Cal stopped breathing. In the intense activity of preparing for the storm, thoughts of Kathy had sunk into his subconscious. Now they became the most important concern in his shattered world. Frantic attempts at communication by cell phone were fruitless. A quick examination of means of transport revealed that the shed housing his vehicles had disappeared. The truck sat undisturbed; the Corvette was nowhere to be seen. But power lines were down and the driveway was obstructed by several uprooted cottonwoods. And the entire area towards the stream was under water. The Raccoon was in flood stage. "Come Sheriff; we'll have to take the road."

Calvin began to run, skirting tree limbs and splashing through shallow stretches of water. Finally he slowed to a fast jog. This degree of activity was unusual for him—Kathy wouldn't be helped if he had a heart attack. Finally he slowed to a determined walk. Sheriff ran ahead, then stood impatiently waiting for him. They made a final spurt of speed nearing Kathy's drive. Cal had a hopeful thought that Kathy might have remained at University; the house had collapsed into unrecognizable rubble. But he was dismayed to see her Toyota lying on its side against the trunk of an uprooted by cottonwood. Breathless, he tried to shout, "Kathy! Kathy! Are you here? Oh, please! Where are you?"

Sheriff had made a direct route towards what had been the front of the house. There was a faint cry. "Cal? Cal, is that you?'

"Oh Kathy! Where are you? Are you OK?"

"Cal, I'm in the basement. But I'm stuck. I can't get out. I'm not hurt, just trapped."

"Oh, thank God! What can I do?"

"Well, for a start, call Sheriff. He's desperately trying to dig me out. It's hopeless though. During the storm, I was sleeping on a cot near the outside wall. Now the floor above has got my leg pinned against the wall. It doesn't hurt, but I can't move it."

Cal thought, "We need a machine—a front end loader or something to lift that section of floor. But I can't call anyone. Roger has equipment, but that's two miles away. And I don't want to leave Kathy." After a quick examination of the wreckage, he called, "Kathy, the piano is resting on that section of floor. Just hang in there though, I'll figure some way to get you."

At first glance, the situation looked hopeless. "Cal, there's an outside door to the basement a few feet away from me. See if that will help some way."

He found the door; fortunately it opened towards the outside. Apparently the wind had moved the upper stories away from the wall support at some stage in the destruction. The edge of a floor and the supporting joist lay just inside the door. Cal moved quickly to lift it; there was no movement—not a job for any one person, but maybe with a lever.

The outlook was discouraging. The roof had ripped apart and tumbled some yards away. The rafters were twisted and split. But some of the interior woodwork was exposed. He found an eight-foot length of varnished two by four showing fine grain; it had to be oak! Then he managed to kick an 8 by 16 inch concrete block free of the foundation. He had tools! Slipping the block past the joist, he was happy to find sufficient space to place the end of his lever about ten inches past the joist; he had seven feet of leverage.

"Kathy, I'm about to try to give you an inch or so of relief. On the count of three, pull for all you're worth—maybe pray a little too."

The end of the lever was about 6 inches above the ground. Cal stood with shoulders back and spine straight. Knees bent, he clasped fingers under the lever and took up slack. Then with deep breath he began to exhale "Three!" Legs began to straighten. There was a creaking of timber and movement! Seconds later, "I'm out! Cal, I'm out!"

With breath gone, vision blurred, Cal carefully reversed the vertical motion until he could kneel with head resting on the lever. Kathy found him so positioned when she was able to scramble from beneath the house. "Oh, Cal! You're hurt!"

"Just meditating! Wondering if I might have blown a gasket someplace. But, no—I'm OK."

Standing, he clutched Kathy close. "So worried, my sweet. Didn't think I would be able to get you." His kisses moved from cheek, to eyes, to forehead, to cheek. Then finding Kathy with upturned face and lips, he kissed her passionately. Her response came a brief fraction of a second later. The embrace lasted for minutes, kisses exchanged numerous times. Sheriff sat watching, massive head slightly cocked.

Finally, Kathy looked past Calvin to the wreckage. "Oh, my poor house. And my piano has a broken leg!"

"Yes, things are pretty bad. But I have you safe! Anything else can be fixed."

"I know! And your arms are a comfort!"

"Mmm—yes. You feel so good. But I wonder about others—the storm was so vicious. Do you have a radio?"

"It's under the house—and I'm not going back under there! We could try the car radio."

"Kathy, that's your car over there lying against that cottonwood on the ground. It might not be wise to mess with it. The storm seems to be over. Do you see anything that we should try to salvage?"

"Mostly it's buried under the house. But I see some sheet music scattered about—might as well pick it up. Most of it is probably in the next county though."

"Not much else we can do here without heavy equipment. Let's trudge back down to my place. It's an even worse disaster, but I have a surprise for you. Coffee is on me this morning."

Morning After

Arriving at the storm shelter, Calvin explained, "Spent a worrisome night here, but safe with Sheriff. Now I'll warm some water on the camp stove; I'll unwrap a bar of Ivory. We can get some of the storm grunge off our hands and face—be good as new." A few minutes later, "Kathy, you are one gorgeous lady. That has always been obvious to me. But I'm seeing you in a different light this morning. Finding you in danger from the storm has given me an enlightened outlook on life. I hope it isn't a mirage!"

Sensing another passionate embrace, Kathy extended a cautioning hand. "Yes the storm was terribly frightening; made me wonder if the end was near. And you must know, I was extremely grateful that you came to rescue me. It was a relief and a comfort to be in your arms. And you know we've become very special neighbors. But I'm finding this very nice increase in affection just a little confusing. I'm not sure I was quite ready for it."

Calvin's happy look slowly progressed to a solemn expression. "Oh, I understand. And I'm quite aware that such a lovely person wouldn't find a scruffy old geezer like me very romantic. But I'd like to always be a special neighbor."

"Cal, you clean up real good. Come here!"

Kathy stood close with an apologetic expression. "Cal, you'll always be more than just special." Then with hands caressing his head, she kissed him softly.

It was a matter of several loving seconds later that Kathy remarked, "And coffee? You promised!"

"Oh shucks! I seem to have forgotten. All right, I'll make coffee if you will try to get news on my radio."

Finally, from a Kansas City station, they found Blue Planet Radio. They had a major disaster to report. "Last night's storms completely devastated an area 200 miles long with a 20 mile wide path. There does not appear to be a single undamaged structure. Thirteen dead are confirmed, but deaths are expected to rise. Casualties are generally heavy although many families took shelter in safe rooms or basement shelters. The mobile home community of Maynard was blown away, but all were safe in the community's large concrete safe house. Governor Jameson has called out the National Guard to prevent looting."

"Well, so much for expecting help. We're pretty much on our own. I have food and water here. Not much variety though." Coffee water had started to boil; he poured it into the pot to drip through the grounds. "Sorry, no cream."

"OK. I'm good for black this morning!"

"I'll make breakfast while we think about how to proceed. I'm pretty good with scrambled eggs. My toast technique on this camp stove is a little iffy though."

"I can cope with burnt toast today!"

After the eggs and toast with large amounts of corrective grape jelly, Cal remarked, "More coffee? I'm not quite ready to face an examination of my house wreckage."

"Looks like the second story with a bunch of furniture is scattered all the way out to the road."

"Yes, and the first floor is tilted up on end. My library and all the electronics must have blown away—probably into the river. That's where the Corvette is. I can see the wheels sticking out of the water. Shit!! Oh, sorry."

"Yes, I know. But look here! There's a whole pile of stuff on the ground—looks like your first floor furniture just slid off when the floor lifted. It's kind of wet though."

"And the plasma TV is in two pieces. I'll have double vision! No great loss. But there's my laptop!" He scrambled to retrieve it. He sat on an unrecognizable section of wall to release the cover and press the power switch. "Say now! We have a miracle. Microsoft and all!" There was another unspoken reason to be thankful. There was now even more reason and urgency to continue with his critical Internet comments on climate change. "Now I wonder if my truck will start. Then we can become part of the outside world again."

The engine came to life with a roar— no problem, tires even had air! "Well now, we can go merrily on our way—almost 50 feet to that tree across the drive!" Then as Calvin examined the shed he added, "We're lucky though. The storm left all my tools here on the bench and shelves. That little chain saw will help me carve a path out of here. If we can get to town, we can improve on our stock of food. And hopefully we can get some tarps to protect exposed things like your piano. And the camp stove and the truck will need some more fuel."

"Any hope for my car? My university students will be hoping to see me next week."

"Tarps first. Then we'll see. I suppose it's proper to get the insurance people to look at the damage before we try to do much recovery. Hope the cell phones work soon!"

It was mid-afternoon when the drive was finally clear of debris. "Kathy, you're getting all scratched up. And I don't know about you, but I'm pooped. Let's relax for a bit. And I believe someone was thoughtful enough to stash part of a bottle of Scotch in the storm shelter. Would you care to join me in chasing our cares away? And a luscious dinner of pork and beans later." With resigned look, he added, "No expectation of affectionate reward later, of course."

"I could use that drink! But Cal, enough of the pretence of a suffering male. I know you better than that!"

Halfway through the Scotch refill, Kathy braved the obvious consideration. "Cal, I have no place to sleep. Eleanor and David live on the other side of town. They might not have had damage. I'm sure she would put me up for awhile. Could we drive over there?"

"Kathy, we can check on that possibility. You would be more comfortable than here and no concern for privacy. I think you may not be surprised to know that I've given the matter some thought. The possibilities here are obviously not designed for personal comfort. There is the cot in the shelter, but it's pretty dismal. But, it happens that the camping gear out by the truck has survived. There are two small tents and sleeping bags. Except for another storm, that wouldn't be too bad. Finish your drink though and we will give Eleanor's a try."

Kathy sat motionless and silent for several minutes. "Cal, we're in this together; I don't want to desert you. Let's continue to work on getting our lives back together here. A tent sounds good to me." Smiling, "An awfully lot better than stuck in my basement! Now do you have a can opener?"

Next Day

Saturday morning dawned clear and bright. Calvin had the coffee made and cups waiting when Kathy stumbled out of her tent. "Pancakes this morning—like it says on the box, with powdered milk. And you can dilute your coffee with that if you're desperate. Now I know you slept safe; Sheriff was right by your tent all night. But you don't look real happy. A few rocks in the wrong places?"

"Cal, later. Behind that brush pile OK? I have to go—now!!"

"Oh, yes. Sorry! Sheriff, stay!"

On Kathy's return, Cal explained, "I've warmed some water. And there's the soap. I found some toothpaste, but we will have to share a toothbrush—you first. Welcome back to civilization, sort of."

"Oh, Cal! I'm a mess. Not used to this sort of thing!"

"You still look gorgeous to me! Now sit down and drink your coffee like a good girl!"Cal's banter could only improve the outlook and soon Kathy was aggressively attacking the pancakes with an egg on the side. "Your second cup of coffee will be ready in minutes."

"Now my plan for today is to go into town first thing for supplies. I'll look around for tarps while you make improvements on our food supply."

"And I have to get a change of clothes too!"

But two miles down the road they encountered a barricade of five cottonwoods lying across the road. Things were not so bad though. Jake Abel from the Mobil station along with Pastor Swenson and his wife Adele hailed them from across the barricade. "Worried about you. Came to check, but trees stopped us. Are you all right? Come on around; walk in the field past the trees."

Cal answered first, "Houses are gone, but no injuries here. How are things in town?"

"Minor damage—some trees. But how did you survive?"

Kathy gave an excited description of her rescue, then a careful description of the food and accommodations at Cal's place.

Calvin added, "We're in sore need of some tarps and some supplies. Then after we get the insurance people to check us out, we'll need some heavy equipment to clean up the mess and try to recover some things. I wonder—could you shuttle us into town and back?" Then glancing at Kathy, "and I suspect that my neighbor may have some other needs that require attention."

Adele looked askance at Kathy, then at Calvin, and back to Kathy. "That sounds awfully uncivilized for you to sleep in a tent and all. We can

put you up for as long as you need, and shop in town for unmentionables.
No reason to live like an Indian just 'cause the wind blew your house
down."

"Thank you Adele, I would like to buy some clean undies and the
like. It's true that the storm has blown our civilization apart, but Cal and
I have to support each other as we cope with the damage. We're neighbors; a little wind isn't going to change that!"

Pastor Swenson chuckled and remarked, "Well, we may learn a
thing or two, but we're not getting any younger. Let's get the show on the
road!"

Shopping was a slow process. Adel insisted on being company for
Kathy. This meant an explanation of the storm damage at every stop.
Storekeepers were helpful however, with suggestions of useful items that
would ordinarily be found in every household—before tornado, that is.
Calvin's visit at the hardware was a constructive affair, netting tarps and
twine, and a promise of prompt delivery of any necessary tools. "On the
house, Cal; bad luck deserves our help. And I know if we get one of those
storms here in town, you'll be ready to give a hand." Then at Perkins'
warehouse, an assortment of cardboard storage boxes and wrapping tape
was made ready for transport. "Just part of the service, Cal. Hope you'll
find enough belongings to fill them. And we'll take good care of Kathy's
piano. Don't worry about the bill; we'll work something out later." Pastor
Swenson had a church member with a van lined up to transfer supplies to
Calvin's truck; another parishioner was busy dismembering the last cottonwood with a chainsaw as they arrived.

On returning to the truck, Cal suggested, "Let's take one of these
tarps up to your place and cover the piano. Then see if we can do something for your Toyota. We can get it back on its wheels and see if it's dri-
vable."

Kathy added, "I parked it just about where it is lying. Looks like
it was just tilted up and wedged against that tree."

At Kathy's, Cal explained, "I think I will not do any more muscle
tricks. We can use the truck and a rope to roll it upright. Maybe another
rope around that tree to belay it so won't blow the tires. You get to drive
the truck."

A few minutes later, "Success! Should have made a movie—
"Dummies Guide to Toyota Rescue". Best check the oil and water—don't
see any puddles or smell any gas though. If you can find a key, we'll bring
the old girl to life—maybe you can drive it to University later."

"The spare key is under the rear bumper. But stop being so familiar with my car—it isn't proper!"

After some minutes of careful inspection by Cal, Kathy asked for the key. "She will behave better if she feels my touch. Stand well away!" In fact, the car responded nicely. After belching a fair amount of blue smoke, the Toyota idled quietly. "She's happy. Sounds like new—or at least like before the storm. She seems none the worse for the wind and your manhandling."

Cal was tempted to say that he would just have to let the air out of the tires to keep control over his houseguest. But the joking atmosphere had turned to somber. The situation had returned to normal insofar as Kathy's freedom to travel was concerned. She could resume her university schedule. And both were aware that she would sensibly adopt the more permanent living accommodations at University.

Insurance

On Monday morning, they were pleasantly surprised to discover that the cell phone signals had been restored. Efforts to arrange for insurance adjustors had mixed success. Kathy was promised a visit in a matter of days, and for a total loss situation she could expect a nominal check at the post office to begin recovery operations. Cal was unable to make contact with anyone at his insurance company.

Kathy was ready to leave for University early Tuesday morning. "I'll get there in time for a shower before class. Cleanliness before Beethoven, you might say. I'll be in touch by phone; leave you a number too. But Cal, I will return—sometime Thursday evening. Make sure my tent is ready!"

"I will wait up—might even arrange a scotch nightcap for you! And if your adjustor shows, I'll begin arrangements for cleanup. We can get your piano and other salvageable things in storage over at Masonville."

"Cal, just let me say that I am terribly glad to have you for a neighbor." And there was a tender, but brief kiss for Cal's anxious lips.

Calvin was finally able to cross paths with his adjustor at the gas station in town. "Look, it will only take you a few minutes to see that my place is totaled. Then I can begin cleanup. And I'll buy you a beer here in town after we return."

It was over the beer that the adjustor confided, "You know, my company has pretty limited coverage area—mostly in this region actually. The storm has hit us pretty hard. There is a rumor about the cash flow situation. Can't promise anything about a fast settlement; you'll just have to proceed with cleanup at your own pace."

"Well, I hope you realize that global warming is likely to aggravate the future storm situation. Hope your company and their backup has had some good scientific advice. The government authorities haven't made much use of the 'precautionary principle' in controlling CO_2. That sort of leaves insurance companies holding the bag!"

Kathy's insurance adjustor stopped at Cal's place Wednesday at midmorning. He agreed that 'total' was a proper description of the damage and promised to call Kathy with more information. "And it would certainly be appropriate to try to rescue her personal things. The piano is obviously damaged; we won't haggle about costs. Get things under a tarp, or better, into a temporary storage. We'll be taking care of those expenses. Kathy's husband was a good person in our agency. We owe him and we'll take good care of Kathy."

Cal replied, "Got my fingers crossed about my insurance. Not much choice now though. Time to get to work on salvage."

Chapter Four

RECOVERY

Salvage

Calvin got a sympathetic hearing from Bill Baldridge, the building contractor in Masonville. "The crane we are using on the supermarket job should do the trick for you. It's busy during the day of course, but we could send it down your way this evening; standard rates, no overtime. An hour's work should move the piano and some of the large debris. Then you can rescue some of the smaller things and get them under cover. You might want to contact Perkins Movers to have the stuff moved to their warehouse."

Cal's investigation of the pile of debris at his house revealed that his electronics had suffered scratches and dents along with severed connectors. Water damage appeared to be minimal; given the chance to dry out, they just might actually work. Paperback books were a soggy mess. He called the cold storage locker in town to reserve space. The mold would stop growing and the frozen water would slowly evaporate. It might be worth a try. After rigging a tarp as an awning in front of the storm shelter, he began a search for usable furniture—a table and a few chairs would be helpful.

That evening, Baldridge's crane gently moved Kathy's piano to an open area where Calvin scrounged some loose siding for a protective pedestal and other timbers to substitute for the broken leg. To Calvin's untrained ear, the C-major chord sounded just fine. Then some not-so-gentle lifting of large sections of house walls and floors exposed a jumble of broken furniture, files, and soggy books, papers, and Kathy's clothing. After a similar exercise at Cal's place, the operator was delighted to share with Cal most of a six-pack of Coors that appeared in the residue under the wreckage.

Calvin's thoughts about his options for the future were full of uncertainties. But there was one worthwhile project that he did not want to abandon. After all, his long professional life need not end just because of a little old windstorm. Hopefully, there was an inquisitive audience out there waiting to hear words of wisdom about climate change. The information was stored somewhere in his brain—must get his laptop back on line!"

It's Later Than You Think!

"Sorry folks; got distracted by a little heavy weather. Hope you are ready for my explanations. It's a little complicated so we'll look at it a few bits at a time. We've discussed the formation of hurricanes and the conditions for tornado activity. When those extremes of weather are brought together, there's hell to pay. How did this happen? Could it have been prevented?"

"To the first question, anytime that a large amount of stored energy in atmospheric water vapor is suddenly released there will be destruction and potential loss of life. The trigger for this energy release is the atmospheric instability due to surface heating and an outbreak of a cold air mass from the polar regions. The recent situation was the result of related atmospheric circulations reacting with maximum efficiency. Consider a comparable situation with, say, 2 inches of rain over a region 10 miles square in 6 hours. It is a straightforward calculation to find the energy release in condensation. This energy was stored by evaporation from the ocean over a period of perhaps at least a week, say, 180 hours. Dividing by that time interval gives the power required for storage. Releasing that energy in condensation in only 6 hours generates 30 times as much power. This is roughly 10% of the electrical power generating capacity on the earth!"

As we have learned, such powerful storms deserve our respect. The energy stored in the earth's atmosphere was accumulated very slowly, using a very small fraction of the considerable power of the solar radiation. It is this explosion of energy in a brief time interval that must be dissipated by the wind and causes the damage."

"Once the mechanisms for transport of the excess solar energy in the tropics are set into motion, there is no human power on earth that can reverse the process. Future mitigation of storm fatalities will require early warning systems and availability of extremely well constructed safe shelters. Reduction of property damage will require costly design and reconstruction procedures. It is far too late to halt the environmental response to our reluctance to take a precautionary approach to the dangers of the greenhouse effect."

Yours for analytic thought,

Chris

Progress

Calvin's efforts for Thursday were initially concentrated on collecting the few undamaged items of furniture and kitchenware for storage under tarpaulins. Soggy items were gingerly assembled for early transport to cold storage. Kathy called at five. "I'll see you about seven. I'll need a drink before we start on the pork and beans. Give Sheriff a pat for me." But did she add, "Love you!?"

On arrival she gave Cal a quick hug before extending a hand for the expected drink. "What, no ice? No reward in Heaven for you! But I'll forgive you this time if you've been kind to my piano. How is my baby?"

"Good as new, I hope. It's got the notes for 'Twinkle, Twinkle Little Star' all scattered about though—you will have to fix that. Maybe we can get it moved to storage tomorrow."

Calvin's dinner preparation was only a slight improvement over the expected pork and beans. "Terribly sorry. Poor planning. I should have gone to town earlier for some fresh meat and vegetables. You are in charge for tomorrow. I'll be happy to cook the steaks!"

****

A search for an AM signal on the battery powered radio finally yielded Blue Planet Radio. It was devoted almost exclusively to the heavy weather in the country's midsection. After several interviews with victims, the attention switched abruptly to a forecast. "From the NOAA violent weather office:" "As you know, our weather systems have a natural period of about a week. In fact, the situation is forming up to be quite similar to last week. However, the deep trough in the jet stream has moved slightly eastward. Last week's devastated area will experience cold westerly winds, but little precipitation. Another moist tropical air mass is moving northward into Tennessee. While there is less moisture available than in the tropical storm of last week, the probability of strong thunderstorms and tornado activity in that region is high. A deep low center is expected to form in Arkansas and move eastward, crossing the Appalachians in northern Georgia and moving up the coast, bringing heavy rains and strong winds to the eastern seaboard. This is somewhat unusual; most of our spring storms travel up the mountains all the way into New England. This storm will intensify as it moves into the North Atlantic. Shipping interests have been notified, and European forecasters have been alerted. Tomorrow's local forecast:

For eastern Colorado and Kansas, southwest winds 10 to 20 miles per hour switching to 23 to 30 miles per hour westerly with gusts to 50 after midnight. Abrupt temperature drop at least 20 degrees. Minimum temperature of 41 degrees in early morning, maximum of 53 degrees in early afternoon. Then partly cloudy skies with winds westerly at 20 to 30 miles per hour. Humidity low, frost unlikely. For Saturday, continued windy with blowing dust; slight warming trend.

For northern Arkansas and Mississippi, Tomorrows temperature minimum of 58 degrees, maximum of 79 degrees. Southwest winds of 15 to 25 miles per hour, skies becoming overcast late in the day. Probability of nighttime precipitation 90% with strong thunderstorms, winds switching to westerly at 30 to 40 miles per hour. A tornado watch may be issued in the evening; be prepared to seek shelter. For Saturday, continued cool with low overcast and brisk westerly winds.

"And now an interview with our climate consultant, Dr. Andrew Fitzwilliams. Andrew, isn't this violent weather rather unusual?"

"Springtime weather in the tornado belt is often violent. However, the activity is rather early this year and especially devastating. You will note that the supply of tropical moisture in the southern states is very high; there is a very large amount of stored energy there. And the amplitude of the atmospheric wave is exceptionally large; it is bringing polar air very far south with strong westerly winds. The situation is perfect for tornados."

"Why is it so extreme this year? Is it global warming and climate change?"

"It is difficult to demonstrate cause and effect in a specific case. However, the high sea surface temperatures are definitely responsible for the increased tropical moisture. Now the North Atlantic has also shown a periodic variation in sea surface temperatures; we're in the warm phase just now. And the current Southern Oscillation of the Pacific has easterly winds in the tropics; that also favors hurricane growth. Global warming may not be the primary cause, but it would certainly aggravate that situation. Folks in the middle of the country are taking more than their share of a beating just now."

"But is this temporary, or will it get worse?"

"If we are seeing a symptom of long term global warming, the prospects are not good. Increased greenhouse trapping of radiation in the tropics will require more active transport of the energy to high latitudes. One consequently expects deeper polar outbreaks and more violent weather. The effects will vary locally of course."

"And remember, about one third of the northward energy transport is carried by the warm Gulf Stream. Then in the Arctic the cold salt water sinks to the ocean bottom for the return ocean circulation; it's called the thermohaline circulation. If that ever fails, the atmosphere will have to take up the slack. Batten down the hatches!"

"Oh my! Hope that doesn't happen. On that note, I see that the Administration's point person for natural disasters, Assistant Secretary Alice Montgomery, has requested time to present the government's view of the situation and to reassure that Federal Assistance will be forthcoming after a committee consideration next month."

"Thank you so much! First, let me ask our citizens to remain calm. We have had these situations before and coped with the problems in admirable fashion. I believe the last problem of this magnitude occurred during a democratic administration some years ago. But this time, the government resources are on call and can be marshaled in a matter of hours. Now as for this rumor of global warming. We expect to receive the final report of our Scientific Advisory Committee on Geography, Weather, and Time Standards any day now. These advisors have been selected from applicants from all interested political parties and religious organizations. They have passed all our rigorous moral and security tests. Now, we have interpreted the Interim Report to indicate significant uncertainties in the temperature records and suspicious bias in scientific methods. Clearly there is no urgent need to depart from our current policies of climate control. More research is needed. Congress has agreed to consider diverting additional appropriations from certain entitlement programs for this task. I thank you for your kind attention. God Bless the United States of America!"

Kathy exclaimed, "Well now! We have our choice of predictions. Our government officials are very comforting; can't wait for them to come hold my hand. Somehow, our little storm doesn't fit the 'not to worry' prediction. But Cal, is it really going to get worse?"

"Well, the prediction for tomorrow agrees with my father's Air Corps 'single station forecast'. Back in those days they were frequently on their own—no network of observations or fancy computers. This southwest wind suggests a steep pressure gradient—no cloud deck though; probably a dry frontal passage. And it will be chilly in the morning."

"I'm not really up on steep pressure gradients, but I understand chilly in the morning. I've read someplace about three-dog nights. But I only have Sheriff."

"Well, aside from the obvious, I suggest that we move your sleeping bag to the cot in the storm shelter. No starry skies, but you'll be warm."

But at 2 AM there was a cold gust of wind and the air was full of dust and debris. Sheriff erupted with a protesting bark. Kathy awoke to look outside; her tent had disappeared and Cal was threshing about inside his flattened shelter. "Cal! Cal! Bring your sleeping bag in here. There's room—on the floor."

These sleeping arrangements persisted for the next couple of nights. Daytime activities of collecting personal items from the wreckage were depressing. The weather added to the discomfort. The wind was cold and the dust was an additional aggravation. Cal explained, "There's been very little rain west of here this spring and no snow since November. And except for the Arkansas, there has been little irrigation water. The Ogallala is low and the farmers have stopped pumping. The topsoil is very dry and getting kicked up by the wind. We're approaching dust bowl conditions again."

By Saturday, they had arranged their meager possessions in neat piles for transport to storage. "Kathy, let's play civilized and have dinner at Wanda's cafe in Masonville. It will be warm and clean and the food is excellent. My treat of course. And to do it up right, I'll let you drive us in your car."

"Sheriff will be lonely here without us. But there will be a doggy bag, big guy. I promise!"

On Sunday, after a leisurely breakfast, they sat out in the sun with coffee. Temperatures had moderated, the dust had settled, and the sky was azure blue. "Cal, it's been good to have your help and company with this mess. It was so discouraging; I couldn't have done it alone. And your hotel accommodations were unique. I will always remember!"

"Sorry about the hot water situation. Would have liked to demonstrate my bath service. Very cozy! My round washtubs are a tight fit."

"Ah yes. I could do with a shower, but can probably wait. I will drive back to University tomorrow, I guess you know. Not sure when I'll see you next though."

"I know. Kathy, have you made plans? Like, are you going to rebuild? Or settle down in that comfortable place at University? That would make sense of course."

She replied, "Yes, the atmosphere here won't be the same without the old house. Now the insurance would easily take care of moving a mobile home in here. But a prefab wouldn't be the same at all. What will you do?"

"Well, first off, I'm not sure my insurance is going to work for rebuilding. There may not be any money. And, like you it wouldn't be the same. And I could move back to Jamestown, but that's not attractive. And without the insurance, I would have to skimp. I may just hunker down here. But Damn! It won't be any good without you!"

"Cal, it's never going to be like it was. But, stay in touch. Don't do anything rash. Let's think about things for a bit. Obviously, life here looks rather dreary for the future. We might do better to leave it." But she thought, "On the other hand…"

Cleanup

Monday, Cal made phone arrangements for Perkins to move things to warehouse. Then, another call to get on a list for a backhoe to come pile up the debris. The call to check on insurance got an answering machine. The warehouse job occurred Monday afternoon; the backhoe would arrive Tuesday morning. The insurance call Tuesday morning failed to get even an answering machine. Cal had a thought that there was a higher probability of finding things of value buried under the wreckage of Kathy's house and they began work there early Tuesday. The cleanup progressed slowly. The work was dirty and depressing. While Cal stood by to rescue any personal items that had been buried out of sight, he felt uncomfortable in retrieving Kathy's toiletries and clothing. And he salvaged any lumber that might be useful in the future for reconstruction. This was done with no thoughts of the future; just seemed to make sense. Finally, the basement area was emptied; he stared for some minutes at the spot where Kathy had been trapped. He shivered at the memory. He thought briefly of saving the oak 2 x 4 and concrete block to share with Kathy, but decided against reminding her of the episode.

The Wednesday work at Calvin's house was a replay of that on Tuesday. The work was relatively brief however; much of the second story had just blown away. Mike remarked, "I'd say you were damned lucky to have that storm shelter; might have found your scrawny body under this mess otherwise."

Cal answered, "You seem more complimentary about my old root cellar than what's left of Calvin! Bet you've never seen one of these. Come take a look-see."

"Not too fancy, but a tornado will never take it. Maybe you should expand it, make it a little more livable before the next one!"

"Suppose? That idea has actually been lurking in the back of my mind. Would you consider carving a bigger cave out of that hillside at your standard hourly rate?"

"Looks like fun! Probably just a little black dirt and a lot of gravel. Then you could box it in and I'll pile the dirt back on top. Add a bit of grass seed and pray for rain. Then you can sit inside and listen to the wind blow!"

Later, after a Chunky can of chicken and rice and three fingers of scotch, Cal considered his options. Living in an expanded cave made as much sense as anything else. Going back to a small apartment in Jamestown would be a dreary way to exist. Of course he could resume activities at the college; the library would occupy much of his time. He had few close friends though; there would be little social activity. Not a great concern though. Life had never been very exciting after Mary's death. Still, the brief period of casual visits here with neighbor Kathy had been extremely pleasant; he had almost begun to take the continued relationship for granted. Of course, the storm had ended all that so abruptly. Now there was little expectation that she would want to return; nothing of value left here for her. It would be good to hear her voice again—and of course that was possible. But he shouldn't telephone. She was probably preparing for tomorrow's class. Maybe just a few minutes wouldn't hurt. He began to punch in the numbers, then almost stopped. She could be out to dinner or a party with other friends. He continued, holding his breath just a little.

"Cal, is it really you? Caller ID looks right—I was wondering if I should give you a call. Didn't want to interrupt and give you reason to burn a Chunky something or other out of a can. Are you there?"

"Kathy! Yes, I survived the monster storm and now it's so quiet that I convinced myself that I needed to hear your voice for a reality check. How are you? Hope I'm not interrupting something there. I just wanted to ask if you wanted me to box the remains of your house and send it to you. Sorry, bad joke!"

"Oh, Cal! You've been working on the mess! Hope you're finished with it. But I don't want to hear about it. Tell me something pleasant. How is Sheriff?"

"He misses you—keeps sniffing about the storm cellar searching for you. Let's see. Something pleasant. Yes! Your piano and everything that didn't blow away or get wet and buried is in the warehouse. Perkins will send you a pleasant little note about it, I'm sure. But you must have a pile of insurance money to take care of that."

"Cal, the insurance situation isn't real good. They've given me some cash to take care of immediate expenses, but they are being very slow about the rest. They say to expect an initial payment in about a month. Then the rest is being postponed for later. They tell me it's a cash flow problem—lots of claims being processed after the storm. What have you heard about yours?"

"Nothing! Absolutely nothing. I can't even make contact with their office."

"I was afraid of that. Cal, my people have told me of a rumor that several companies, including yours, have defaulted on payments. The storm was just too much for the insurance industry. I'm so sorry. I wish the news were better."

"Kathy, I'm getting used to bad news. But I'll muddle along. Now, my turn; tell me something pleasant. Like how are your classes going?"

"Well, actually I have one young tenor who is really promising. He will probably move on to Julliard. And my class is about to start a study of the career of Kiri Te Kanawa. Do you know of her?"

"Not a lot. But I have some CDs she has done of Cole Porter, Jerome Kern, and others.'

"She's opera, Cal. Maybe you have the one of Irish songs, and another of Maori songs? Interesting, because they were her parents."

"Sounds like I should come to your class!"

"I'd like that—but I know you've got other problems just now. Like, what are you going to do without insurance?"

"Might just stay here and live in my sod house. Can't expect a visit from my neighbor if I do that though. Can you think of a way that I could solve that problem?"

"Cal, it's not just your problem you know! Keep your shirt on— I'm thinking about it. Seriously!"

"My God, Kathy. Are you saying I should be optimistic? No way that I deserve that, I know!"

"Now listen to me. This summer term ends next month. Maybe we could get together to talk about our options—finances and such. If you would invite me, that is."

"I'll send you an engraved invitation. And find you a civilized bed in a motel somewhere near."

"Cal, it's only a bit over an hour's drive. I'll be down early for a short visit during the day. Don't worry about a motel!" Kathy thought, "But I'm not driving back after we have dinner."

With the lull in salvage activity, Cal gave further thought to his ghost-writing project; time to get to the heart of the global warming matter.

It's Later Than You Think!

Carbon dioxide is a minor constituent in our atmosphere. By itself, it does no harm to humans or other earth's creatures. In fact, we exhale CO_2 as we breathe the oxygen, which our body needs. Now, some legal eagles have insisted that it is not a pollutant; it doesn't make us sick. Obviously though, if some disaster replaces oxygen with carbon dioxide we will be in dire distress. There are places in our solar system that have that problem, but it's not our concern—yet. We have a different little problem that is an insidious threat to us.

Carbon Dioxide, A Greenhouse Gas

You will recall that plants have used photosynthesis to grow, using CO_2 and generating O_2 as a byproduct. We might hope that process will always keep our atmosphere in balance, wishful thinking perhaps. But what's the problem about a wee bit more of this so-called non-pollutant?

Well, there's lots of talk about carbon dioxide and global warming. Just false claims and scare tactics by environmentalists? Best start at the beginning then. Measurements of its infrared absorption became available a very long time ago. Shortly thereafter, concern was voiced that its greenhouse behavior posed an eventual threat to climate change. So about 50 years ago a campaign was initiated to make atmospheric measurements of its concentration from the top of Mauna Loa on the big island of Hawaii. That choice of location precludes influence of power plants, forest fires, or other local effects. This database represents a global record of the general CO_2 abundance and continues to the present time. So there's not the slightest doubt about the recent history of the CO_2 increase. The relative amount of CO_2 has increased from a historical baseline fraction of 280 parts per million to 315 parts per million at the beginning of the measurement series, to a value of 360 parts per million in 2000. That is, the amount has increased by about 1/3 during the industrial era. The sensitivity of the measurements is such that the seasonal balance of production in winter and loss to plant growth in summer is readily visible. The rate of increase continues to steepen. And the atmospheric scientists with their fast computers tell us that the enhanced greenhouse effect attributed to this CO_2 increase has produced a global average temperature increase of 0.6 degrees Celsius (a bit over 1 degree Fahrenheit).

This information has stimulated efforts to determine the earlier history of atmospheric CO_2. The snowfall forming the glaciers of Greenland and Antarctica has trapped air with minute samples of CO_2 of that period. Other measurements of the glacial ice furnish a seasonal calendar. These records show that present day CO_2 abundances are higher than at any time in the last 600,000 years!

Where does this extra CO_2 come from? Burning any carbon containing material produces CO_2, things like candles, methane, wood, coal, and oil. It doesn't take much imagination to see that the increase of CO_2 in the atmosphere during the recent industrial era comes from the burning of coal and oil fossil fuels. In fact, calculations based on the consumption of coal and oil yield increasing CO_2 amounts that parallel the atmospheric observations. These calculations of industrial activity are completely independent of the Mauna Loa observations; they even reflect the perturbations in industrial productivity during wartime years.

Now, by virtue of our population and technology, we are burning coal and oil at an increasing rate—more and more CO_2. So, how long before we have doubled CO_2? And what will be the temperature increase? Your scientists know—listen to them!

Yours for analytic thought,

Chris

Reconstruction

Calvin was actually somewhat relieved with the news of the insurance disaster. There might eventually be some fractional adjustment, but there was no possibility of rebuilding to the level of previous accommodation. Besides, there was no guarantee that it wouldn't be destroyed by another storm—and more uncertainty about insurance. And there would be no plush cushion of cash to live at Jamestown with frequent trips to Hawaii.

He began to speculate about his 'sod house'. One could design it to be comfortable with modern furniture and appliances. Perhaps it could be done within the limits of his invested savings. And it damn well wouldn't blow away! A visit with a contractor, maybe Abner Hansen, a local acquaintance, might be entertaining.

It was late Thursday afternoon when he made contact with Abner. They arranged to meet at the Cattlemen's Bar to talk about the possibilities over a few beers. "Abner, I've got one of those root cellars like in the old days. It was always available as a storm shelter, and I survived our little windstorm there. Now I'm not much tempted to put up a new house to get blown away again, especially with questionable insurance coverage. I got to wondering about a more elaborate underground place—concrete walls with strong windows and extra lighting, maybe about 1000 to 1500 square feet. Use salvaged lumber for roof and internal walls. Suppose it could be done without breaking the bank?"

"Well, let's see. That would be reinforced concrete for exposed external wall—don't want it to have it fall apart with a gust of wind. Probably concrete block for underground walls. Windows would be special; you might still want to have shutters during a good blow. The excavation ought to be rather straightforward. Wouldn't take long to work up a fair estimate of that. Then the inside work would depend on how fancy you wanted to be. "

"Tell you what. I'll work up a sketch to give you a start on the external. Haven't thought too much about the inside. Anyway, I would plan to do most of that work. I 'spect I could learn to do the plumbing and electric, maybe with some expert advice; and inspections too of course. Not as if I had lots of other things to do like go to the office to make my next million. Then just do it until the cash runs out!"

Cal imagined the old root cellar to be the new bedroom, about fifteen feet by fifteen feet. Another adjacent room, about equal size would be his study-library-computer area. Then a kitchen-dining-work area about fifteen by twenty-five feet with large outside entrance door. Rudimentary shower and toilet facility in the corner with entrance from either kitchen or study; small laundry facility with clothesline outside; human dishwasher at the sink. And beyond the kitchen, an equally spacious room and outside entrance for whatever. An old style 'sitting room' or a spare bedroom, perhaps for Kathy? Wishful thinking!

Decision

Abner's estimate ran a bit over a third of Cal's savings; his experience suggested the internals would be roughly the same. Decision time! There was a vague thought that Kathy ought to be consulted. Silliness! The prospect that she might ever consider a future abode with him was off the wall. Still, ought to call her to clear the air. She might have some other suggestions that ought to be considered.

"Kathy, farmer Cal calling. Just finished with the milking and thought to chew the fat with you before I put the chickens to bed. Got a few minutes?"

"I've been sitting on a low note, hoping you might call. You've made my day! Actually, I have an evening recital to attend in a few minutes. But, what's up?"

"Nothing much. Just had to check that I had a clear channel to your voice. Insurance people have left me high and dry; I'm not going anywhere. Been thinking about making some slight improvements to my sod house—winterize it don't you know! Actually, make it a plush job for my golden years. Whenever you drop by, I'll be right here. Any inspiring messages from you?"

"The sun will come up tomorrow. That's all I know for sure. But now that you ask, if you can keep the weather under control I'll bring a picnic basket down with me at the end of the month. We can sit by the river and watch the water run downhill—and try to forecast our future. Cal, I really do have to run—talk to you later!"

"Bring mustard. I'll cook burgers!"

****

Blue Planet Radio on Friday night was dominated by descriptions of heavy precipitation in Alaska and Scandinavia. A series of storms out of both the Pacific and Atlantic had moved into high latitudes, bringing heavy rainfall to coastal areas and continuous snow to higher elevations. Their climate consultant, Dr. Fitzwilliams, had the following comments. "Recent weather patterns have been characterized by intrusions of moist tropical air into northern latitudes and exceptionally deep outbreaks of polar air into the continental North America and Eastern Europe. This is, of course, a somewhat extreme episode of the transport of the absorbed solar energy from the equatorial to polar regions. While the increased moisture is generally beneficial, there has been considerable stream flooding along the coasts. The effects at higher elevations will be long term. Snow is being added to the source regions of the glaciers at an increasing rate. While those glaciers have been retreating in recent years because of melting at low elevations, the addition of heavy snow at their source may reverse that process. Similarly, the increased reflectivity of the snow at these high altitudes may modify or halt the present global warming. Some cause for optimism. But in the meantime, the continual outbreaks of dry polar air have initiated drought conditions in much of the U.S. mid-continent. These observations must be incorporated into existing computer models of global climate. I'll update you on that as the information becomes available."

****

Cal formalized the project with Abner at the Stockmen's with a handshake. Abner allowed as how, "Jack Sorenson will probably get started on the excavation on Monday; should pour concrete sometime the following week. And I know of a couple of young fellows who would rather do this sort of thing than work out on the farm. I'll send them out if you like; they'll want just a tad over minimum wage, they'll do whatever you say— and maybe better quality than your best!"

The excavation turned up some surprises. "Cal, that rock wall in the back of your storm shelter extends some 70 or 80 feet along the hill. Looks like there must have been some kind of structure here in the past— maybe a barn or shed of some sort. It's kind of fancy for that though; rocks have been shaped, no mortar. Wonder where they got them; nothing like that exposed for quite a distance from here. And come to think of it, there's a gap in the wall, kind of a recessed area like a fireplace or some such. Now a barn or cattle shed wouldn't have a fireplace!"

Later, "Cal, look here. Your sketch says to go down another 16 inches. But I've found a level layer of smooth packed clay here that could be a good foundation for a floor. Might not even need concrete."

"You know, that's just like the dirt floor in the root cellar; in fact it's the same level—must be more of the same. Mystery deepens, you might say! No, that's fine; leave it!"

The concrete outfit sent their crew out to lay out the frames and reinforcement. "Reinforced front wall will be tornado proof so to speak! Underground side walls can be concrete block; you can add wallboard or whatever to suit."

Cal altered his floor plan just a bit to accommodate the discovered fireplace in a study-library area. Abner's research had established the dimensions for picture-window size hurricane proof windows for the bedroom, study, and 'spare' room; similar smaller windows would work for either side of the entrance door to the kitchen. "These should arrive in about 10 days."

Tom and Dick

Abner introduced his young helpers as 'Tom' and 'Dick'. Calvin asked, "Where's Harry?" This was met with mutual questioning looks and blank expressions, until Dick replied, "He doesn't like to be part of a crowd." The outlook was good; these guys were going to catch on quickly.

Dick was short and stocky, dark hair and the start of a mustache. Arms and shoulders suggested that he worked out with weights. Tom was tall and lanky, blonde with blue eyes; Scandinavian heritage was not unusual in these parts.

Abner explained, "Tom is a Swedberg from over by Masonville. Dick is from that big Carillo farm family; their farm is just north of here a ways. Cal, these fellows are here to do whatever you say. They can't claim to be experts so you will have to ride herd on the results. Still, the farm boys around here have lots of hands-on experience. I figure the three of you can manage to get the jobs done; trial and error maybe. They're husky lads though; nothing too big for them. Put 'em to work!"

Cal remarked, "Yes, they look to be pretty healthy specimens. Now guys, you can see that I'm a bit past my prime—had a sheltered life being a college professor. I have some thoughts on how this finished project should look. We'll have to put our heads together on how to do it. So let's get acquainted. Tell me, what are your plans for when you get kicked out of the nest and on your own."

Tom answered, "Well, we're going to start classes at Jamestown next month. I think Dick wants to have some fun before he has to take over the family farm. Me, I'd like to be a medical doctor—should be an exciting career."

Dick interrupted, "Yeah, Tom is the one looking for excitement. I'm just going to like take some business courses—hope to keep the farm out of the red! Dad's gone, but Mom would like to stay on; can't have her worrying about money though! "

"Well, let's have at it. First thing is to sort through that old lumber for internal walls and roof. Then we can get more precise about dimensions."

The hardwood floors from the old house were ideal for the ceiling structure. The internal walls and roof supports used most of the salvaged wood from the farmhouse. Frantic activity with saws and hammers for a week and they were ready for a change.

Abner recommended 'Frank and Ozzie's Plumbing' for advice and supplies for plumbing. Connections to the old well and septic facility were not terribly convenient, but possible. The project occupied Cal's time and patience for the next week. He began to appreciate the 'exorbitant' plumbers' rates. Next time he would splurge on a professional!

The electric situation was just a bit awkward. There were some embarrassing moments when the linemen discovered Calvin's arrangements for emergency power to his microwave, TV, and computer. "But I didn't bypass the meter; haven't been stealing your 60 cycles!"

Marty and Jim's Hardware Store had a pile of new catalogs on modern light fixtures and super efficient lights. Even with the large south windows, there would be a need for near continuous lighting in the back of the 'cave'. Cal had designed conservatively for eight-foot ceilings; there would be plenty of room for ceiling tiles or whatever to cover his electric conduits sometime later—if the cash held out.

Tom and Dick had been faithful workers, arriving after early breakfast and departing just in time for supper at home. Near the end of the month, they discovered that all necessary systems were 'go'. Dick remarked, "It ain't beautiful, but by golly it works." Tom added, "And it's a big step for Cal, if not for all mankind." Cal agreed, "Couldn't have done it without you fellows! Now tomorrow you can drop by at your leisure for a few finishing touches. Then about four o'clock I believe I can arrange to have steaks on the grill and a few cold beers. Let's celebrate a bit!" Looking ahead to Kathy's visit at the end of summer term, he would have a fair amount of pride in showing it off. It would be nice if he could afford to furnish the place with lots of plush furniture; that would be impressive. But even with that he couldn't expect Kathy to accept a proposal to move in!

The Cave

Calvin's eager young assistants had also begun to take considerable pride in the design and construction of the 'cave'. Just prior to their departure to begin college courses, Tom explained, "Dick and I have enjoyed working with you Cal, and sort of think of this as our project. We'd like to continue working on it. We'll be coming back frequently on weekends and could give you a hand—for free—or maybe for a steak dinner! We think you need carpet for your bedroom and study—tile for the kitchen and maybe the spare room. If you can swing it for the materials, we could give you a fancy floor surface some weekend."

Dick added, "There's a couple other ideas I'd like to work on. You need some reliable ventilation—can't have the place smelling like burnt toast or maybe mildew. Now my uncle in western Kansas has given up on growing wheat out there in the dust. He has abandoned the farm—let the bank worry about what to do with it. He is trucking the machinery and household goods to Canada. With the warmer temperatures up there, the forests are being cleared and they're plowing up the fields for wheat. Won't slow down global warming, but people have to eat. Anyway, I have my eye on a big fan assembly that was used to dry grain; he left it behind. Now if we mount it on your roof and run it slow like, with an inlet here and there, I bet you would always have fresh air."

Cal had a prompt reply, "Sounds real good! The circulation up that old chimney leaves something to be desired. We could probably design a wind proof shelter for the fan on the roof. But you know, it's always a tad cool and damp in the cave. Now I wonder; suppose we could warm up that inlet air just a little?"

Dick scratched his head. "Maybe what you need is a greenhouse with river rocks from the Raccoon for solar storage. Inlet air from there ought to be about the right temperature. You do the engineering and I'll collect the rocks. And what about some wind power? I know where there is an old windmill that isn't being used to pump water anymore; well is dry. I bet that could be adapted to charge batteries to drive your roof fan—maybe some other things too. Don't want you to go hungry just so you can pay your electric bill."

"Say now! Maybe I should also have some solar cells to drive my computer; and with our Kansas sun, I could probably send some juice back to the grid. Wouldn't that be a switch!"

Calvin realized that he had neglected another fundamental atmospheric process; it wasn't strictly physical. Biological processes were generally more complicated, but this one was important in the global warming scheme of things. Best give it some attention. And he wondered, "Maybe it's time for Chris' readers to think about some constructive actions to stave off global warming. Zorro might get them to at least think about planting more trees!"

It's Later Than You Think!

Trees

Did you know that your very existence on this planet depends on the trees in your backyard? Oh, sure. You grow apples, and burn wood in your fireplace. And you don't have to live in sod huts anymore. Ah, but you also need to breathe! The oxygen in the atmosphere is almost entirely dependent on the conversion of CO_2 to O_2 by plant life. Now this process, called photosynthesis, got started on earth somewhat before your time. (We won't argue just now about who designed it.)

Photosynthesis is a long word. Can you spell *it*? Oh, sorry—a joke my mother taught me. A little mysterious though; what does it mean? Sounds like something that Dupont or Parke-Davis does. Actually it's another kind of solar power. The energy carried by solar photons can be used by bacteria and plants to use some simple molecules to make more complicated ones. Bacteria multiply and plants grow. Happily for us, the plants convert CO_2 to O_2 in this process. Photosynthesis uses CO_2 from the atmosphere along with water and nutrients from the soil to grow plant material and release O_2. Photosynthesis is almost entirely responsible for the oxygen in the atmosphere. And the earth's forests sequester a significant amount of this carbon in the wood of the trees.

While the dominant factor in CO_2 increase is the burning of fossil fuels, another contributing factor to the increase of atmospheric CO_2 has been the conversion of forested land to agriculture and other activity of humans. The failure to restock the forests has upset the atmospheric balance. Replanting trees would be a partial solution to our CO_2 problem. A positive step in global efforts to reduce greenhouse gases could credit forested nations in comparable fashion to those who reduce emissions.

Well! It has been observed that increased CO_2 benefits plant growth in controlled situations like greenhouses. Some people then argue that more CO_2 from burning fossil fuels would be a good thing. More oxygen and more food from plants! A predictable consequence is the argument that increased atmospheric CO_2 and warmer climate will be beneficial to agriculture and food production; burning fossil fuels is in fact good.

I wonder now! Life is pretty good the way it is. I wonder if you are wise enough to improve on it. Murphy's Law number umpteen: It's almost certain to be worse! And there are ongoing scientific experiments to find out. Research in progress indicates that the growth benefits are not likely to persist at high CO_2 levels. Plants also need water and fertilizer to use the extra CO_2. (And extra fertilizer makes more of another greenhouse gas [N_2O] in the atmosphere!) Global warming is also pretty likely to mess up the present balance and distribution of water. Crops trying to grow in the desert cannot use increased CO_2. Plants and puppy dogs and humans need water too!

Photosynthesis is unique in the direct use of solar energy to remove CO_2 from the atmosphere. Our engineers have invented other techniques, but these generally use fossil fuel for the energy to build the tools for such processes, or to concentrate and store this substance. But I have an off-the-wall suggestion. Perhaps we could contract the services of an army of the mythical Maxwell's Demons to trap the CO_2 molecules. Physics books describe their imagined ability to use a trap door to select fast molecules from all the rest; they would just have to modify their technique for CO_2 molecules. Then we could make constructive use of this gas to inflate soccer balls and SUV tires. Now I suspect we may soon hear of a proposal from the Vice President of West Slope Coal to secure a million dollar grant from the Department of Energy for trap door design and recruitment of the army. (But I hope that the government bureaucrats will not credit me with this idea!)

Now, there is no arguing the fact that burning fossil fuels is increasing the CO_2 in the atmosphere. But we are aggravating the situation by cutting our trees, and burning the forests is adding insult (CO_2) to injury. Replanting trees would be a partial solution to our CO_2 problem. Trees are a renewable resource for buildings, furniture, and firewood, but only if you replant the trees! And I think I shall never see a concrete condo lovely as a tree! How about tax credits for planting a tree?

Yours for analytic thought,

Chris

Chapter Five

COMMITTMENTS

Picnic

Kathy called during the last week of summer session. "Tell me, is caveman Calvin still wandering about with his dog Sheriff? Haven't seen any smoke signals from him in a few days. Hope he isn't out doing the hunter-gather thing—I'd like to drop by to see him this weekend."

"Oh my yes! I'll just have time to go to town and get the red carpet! Just keep that Toyota between the fences! Seriously, I'm looking forward to seeing you—I've missed you."

When Kathy pulled into the drive on Saturday morning, Sheriff was first to greet her. There were the ear massages and head pats until Cal could close in. "Hey! This old dog needs love too!" But there was only a quick embrace and cheek-to-cheek before Kathy exclaimed, "Cal, what have you done? That storm shelter—it's had a face lift!"

Calvin had a sudden worry that he had been too independent in his plans. True, it was his to manage as he saw fit, but he hoped that he had not done something to annoy Kathy. "It's really not much. Just figured that if I was going to be stuck here, I might as well be more comfortable. And it's not finished—something to keep me busy in my golden years." Then bravely, "Could I give you a tour?"

Kathy's eyes widened as she approached the massive front door and realized the size of the picture windows. And the cave was no longer dark!

"The kitchen with sink and all—just like downtown. The old fridge still works, but I'll get a new one some day soon. I had to buy a new microwave, else I'd starve. Next door here is my study-library. Not many books right now. But most of my electronics still works, like the computer and music system. Speakers are gone of course—will have to replace them eventually. The fireplace sort of came with the excavation, but it needs a new chimney before the weather gets cold. And that's my bedroom; no closets yet, but don't have much left in the way of clothes anyway. Oh, that room on the other side of the kitchen—just got carried away in making a big cave. I'll think of something," then jokingly, "maybe start a bed and breakfast!"

This was a slightly rude surprise to Kathy. Cal had moved ahead with his life. He was a survivor. She became aware that she had adopted an attitude of responsibility for their future. This didn't make much sense. There was little reason to pool their resources—unless—but she hadn't made that commitment yet.

"Cal, you're amazing. I had a thought about buying you a new tent, but you're way ahead of me." Way ahead, and why does it bother me so? "Now, I trust you have the picnic spot by the river all arranged. I have some goodies in a basket and you are cooking the hamburgers. Right?"

"Kathy, we've been apart too long. Poor communication! I changed the menu; it's king salmon filets 'cause I can do it in the microwave. But then when it's cooked, we have to cover it carefully and drive up to your place. You see, the storm has made a little channel like an irrigation ditch just this side of the river here. But there's a big old cottonwood up at your place that survived the storm; we can spread a blanket there in the shade. And if you've brought the wrong color wine, I just happen to have some Schwartze Katz from Germany."

"I agree Cal, about being apart too long; it worries me. I didn't remember your being so much in charge of things. Oh, but I like it—just hadn't noticed that part of you before."

"I don't mind your getting to know me better. Actually, the storm forced me to abandon my casual drifting lifestyle. I have no reason to expect the good things to fall my way at this time of life. Ah, but I continue to hope!" The final comment reassured Kathy; perhaps they weren't so far apart after all.

"So, we'll have to take the truck so that Sheriff can come along. I have a soup bone for him. Otherwise he would be drooling all over our picnic. And I rescued one of my mother's old quilts to spread on the grass. It was a bit smelly when I found it in the debris from the storm. It would never survive a laundry so I just let it air for a few days. Now I won't ask when you had your last blanket party—just hope the food will keep you looking to the future."

The early morning temperatures in late August had been far from balmy, but the afternoon was comfortable with a steady southwest breeze.

The air was dry; the big cottonwood was beginning to shed a few yellow leaves. The tall grass along the river had matured. Seeds were ready for another year; still there was a carpet of green using the soil moisture from the nearby river. But the pastureland across the road was bone dry. Murphy had begun feeding the cows hay, and hoping for fall rains. The weather systems were temporarily on hold. Another polar outbreak was poised to the north, but the warm moist air mass was far to the east. Cal was expecting another cold dusty episode in a few days. This was perfect picnic weather though and he was grateful that this would be a relatively pleasant experience for Kathy. Perhaps it would encourage an outlook for further visits. Life might be a lot less dreary if Kathy were always near.

Some twenty minutes later the picnic supplies were being ferried to the shelter of the cottonwood on the riverbank. "Food first while it's hot. If we can persuade Sheriff to chew on the bone downstream a way, we won't have to listen. If he insists on being sociable, I can always fasten him to another tree with his leash."

"Sounds like a plan. Now if you will just spread yourself comfortable like on the blanket, I will arrange the goodies within arm's reach. But mind your manners. Hope you haven't developed bad habits from living in your cave!"

"Kathy, it's good to have you here. I know this is our first picnic, but it seems so right and natural."

"Yes, and if you will open the wine, everything will be perfect"

The picnic lunch progressed quickly to cheesecake and wine before conversation resumed. Cal and Kathy were both reluctant to begin a serious discussion about future plans. "Cal, you've arranged a perfect day. It's warm and pleasant, not a cloud in the sky. I could sit here forever. Can you make the time stand still?"

"Yes, a full stomach, perfect temperature, and a good view of our river and trees is easy to take. And the good view includes you I must add."

"You say nice things. You and Sheriff are good company."

"Well, including me with Sheriff is reassuring. Speak of the devil though. I think he is finished with the bone and is headed this way. Let's have a wine refill and go for a stroll along the stream."

"Cal, what was it you said about an irrigation ditch down at your place?"

"Well, I hadn't noticed it earlier. But the high water must have opened up the entrance channel in the edge of your property; now there is a continuous stream—not a lot of flow, but enough to get your feet wet. There's not enough fall for it to be a millstream. It certainly wasn't meant to irrigate the pasture or hay field. Maybe it was just for a garden or an orchard; there are actually a few apple trees along the upper end. They might even make apples now that the storm has taken out some of the cottonwoods. I have to wonder if we are about to relive some old history. Nothing in print that I know of, but there is a very old rock wall and that fireplace along the back of my cave. Maybe some farmer from Missouri on his way to California stopped here for a few years. Or maybe some long forgotten Indian tribe had advanced to fireplaces and irrigated farming. More likely this was an effort to subsist during the dust bowl days."

"So why was it abandoned? I'm not much into gardening, but looks like it might work."

"Oh, this was good enough for beans and potatoes and chickens but wouldn't put any money in the bank. Probably the rains came and they were tempted to go back to planting wheat and raising cattle."

"Cal, my wine is gone and my feet hurt. If you take me back to our picnic spot, I might let you hold my hand."

"Kathy, I'm yours for any small reward. And I have arranged for a gorgeous sunset and as a precaution I will maintain this nice dry breeze to keep the mosquitoes away."

On return to the picnic spot, they discovered that the ants had finally discovered the potato chips and the nearly empty dish of baked beans. A long column of foragers was transporting the residue home to the nest.

Kathy exclaimed, "Sorry guys, the picnic's over. Ah, but they missed the wine bottle; all is well. Cal, I'll stack the remains of our picnic in the basket. Then maybe you can shake the ants out of the blanket and move it to a fresh spot."

With replenished wine and a comfortable west-facing slope, Kathy added, "It's going to be a beautiful red sunset; hard to believe we've just had that disaster of a storm. What's your prediction for the future?"

Calvin suppressed his explanation for the red western sky; the wind continued to stir the dust and a new cold dry air mass was on its way. "My meteorology professors always told me to never give a forecast except for a fee. I hope you are about to make an attractive offer."

"Not just this minute. Besides, I've been studying that series of Internet climate discussions by Chris Baldwin, the environmental assistant to Professor Mandryka of Golden West University. He's the expert isn't he?"

Calvin sighed. This subject required some care. "I believe his explanations of the fundamentals are good. How did you happen to discover that website?"

"Oh, everybody knows about it. His discussions make sense; they're dependable. He doesn't predict the end of the world, but he doesn't pull any punches about the science."

That response was gratifying, but Cal had some misgivings. There was a basic response to protect his secret contributions. But there was also a desire to be completely honest with Kathy. Still, the relationship was young; perhaps he need not bare his soul just yet. And the series of web-site contributions was well along to maturity. By the time the readers discovered there was some real authority behind the fictional Chris Baldwin, Cal would have retired from the scene.

Time to change the subject. "So what are your thoughts for the future? Hope they include more days like today."

"Cal, my insurance company says I'm right at the top of their list, whatever that means. And rebuilding something comparable to the old house would never be the same anyway. It would be vulnerable to the storms, not like your cave. Insurance companies don't seem too dependable; besides insurance rates have doubled in this area. I only have enough ready cash in the bank for a mobile home; that doesn't sound terribly interesting."

"Those things blow away in the first excuse for a storm." The conversation dwindled to a depressing lull. "Damn it Kathy! You must know how much I want to help. Life has been a long twilight for me after I lost Mary. It has become interesting and worthwhile since meeting you. But time is being unkind. I'm just an ancient admirer whose house and meager fortune have blown away. It's frustrating as all hell!"

"Cal, you of all people should understand my uncertainty of the future. Albert's illness and passing left me with a dedication to my profession and teaching. A new life and love was out of the question. But Cal, isn't it obvious that my thoughts of returning here are mostly about you? And you're not ancient; you've just lived a bit longer than I; I gave you a head start!"

Kathy's reply was a glimmer of hope for Calvin. "The breeze is getting just a mite chilly. Would you like to sit a bit closer?"

With her head on Cal's shoulder, Kathy wondered aloud, "I suppose one could haul the mobile home away to a safe place during tornado season. Sounds rather silly though; still doesn't make it a palace."

Minutes after bestowing a tender light kiss on the top of Kathy's head, Cal stiffened abruptly and asked, "Have you ever examined one of those big recreation vehicles? Some of the new fifth wheels are expandable; the rooms slide out to become double-wide. And they are designed to be towed; moving would be no problem." A slight pause, then, "And my truck has the power, just need to add a towing platform."

"Sounds expensive though."

"We might get lucky on that. Sometimes new retirees invest in a new one and take off on a trip to Alaska. But then they discover it's more strenuous than they imagined, or there is a health problem or some such. Now they have this albatross sitting out by their garage. It's still in good shape and they might be happy for someone to haul it away at a considerable reduction in price. We could check on the Internet."

"Later; I'm just getting warm and comfortable!"

It was about 30 minutes after sunset when they returned to Cal's cave. "I hope you're not really going to start up the highway. I have a bed reserved for you in an almost new and exceptionally unique abode. The mattress is new and comes with the best of company...." Then observing a tolerant but slightly reproachful look, Cal added, "I've changed the sheets and I have a sleeping bag in the spare room. I'll pop some corn and heat up some soup; then we can sit by the fire and ponder life's little problems."

"Cal, I can see that our approach to these things is very similar. I had no intention of driving home tonight. Sitting by the fire with you is good for now." Kathy's thoughts on progress to other arrangements remained open.

Calvin's makeshift sofa by the fire was an arrangement of cushions rescued from the damaged house. This was covered with an exotic sheepskin blanket from New Zealand. This was a gift from a science colleague in return for travel and accommodation arrangements during a professional visit to the U.S. It had survived the storm in a well-protected storage closet.

Kathy was only slightly surprised to discover the kindling was already prepared for a small fire. The small supper was served on the floor by the fire. Another bottle of Schwartze Katz appeared. Conversation was replaced by lengthy gazes into the flames and intimate thoughts of the companion. Cal replenished the fuel for the small fire and returned to hold Kathy close. Some minutes later he sat forward to look fondly into Kathy's eyes. There was an unspoken question. Kathy smiled slightly, then lowered eyelids briefly and tilted face in welcome. There was an exploratory kiss; it was followed by a series of tender, then passionate promises. Progress was slow and deliberate. The flames in the fireplace flared briefly and subsided to a warm glow.

The new mattress was warmed by its occupants somewhat later. The sleeping bag in the spare room remained neatly secured in its protective cover.

Kathy returned to University in late afternoon of the following day to prepare for classes and music activities in the next term. Calvin was left with a renewed sense of dedication to arranging comfortable and reasonably attractive living accommodations for Kathy at her homestead. While their experiment at physical love was definitely pleasurable and satisfying, there was no sudden dedication to share common toilet facilities. And Cal's cave took some getting used to!

RV Shopping

The possibility to reconstruct a conventional modern house on the site of Kathy's old farmhouse was held in abeyance. Kathy explained that Arnold had made conservative investments that would cover much of the cost. However enjoyable a new house might be, its future value was questionable. While these rural settings were pleasant, a home five miles from Lynn hardly qualified as suburban. Kathy's future options would be tied up in that little house on the prairie.

Kathy had not immediately expressed disgust with the notion of spending weekends in a fancy fifth wheel parked on her property. In fact, that idea had seemed slightly intriguing. And that modest investment would be adequately protected by the option of towing it into safe territory during tornado season. The additional possibility of living with the bears in a National Park had not yet taken root.

Calvin's investigation of the Internet took him into a wilderness of unclassified mobile homes from Florida to California, some on wheels, others anchored to concrete slabs. Progressing to RVs was even more discouraging. Ads for new vehicles ranged from 40 ft diesel powered monsters to popup campers. Previously owned trailers and the like offered 'as is' at $500 to $50,000 could be found from Maine to Oregon. Shopping looked to be formidable. Come lunchtime, he gave Sheriff a boost onto the truck bed and drove into town for a bite at Ben's Cafe. The Masonville Sunday Gazette at the counter headlined the firing of the Broncos coach and the closing of the tractor factory in Pueblo. Comparable disasters, those. But there would be old-fashioned classified ads for things on wheels located within a day's drive. And he could always revert to a study of the colored comic section.

In retrospect, this was a logical step in the right direction. Senior citizens with a need to stabilize their finances would look to understandable and trustworthy methods to communicate. And there it was: 'Immediately available for your trip to Arizona: Fifth wheel, 25 ft expandable, low mileage, excellent condition, tow vehicle optional. Make offer.' There was a telephone number and address in Masonville. Just what the doctor ordered. Cal made the call.

"Hello. This is the Carlsrud residence, Betty speaking."

"Yes. My name is Calvin Carpenter. I'm calling in regard to your ad for the fifth wheel recreation vehicle. Is it still available?"

"Why yes; you are first to call. Would you like to see it?"

"I'm about twenty minutes drive away. Is now a good time?"

"Oh my yes. Later might be a problem; we tend to take a late afternoon nap on Sunday. You have that address; we'll look forward to seeing you."

The Carlsrud house on Elm Street was a large old two-story structure in slight need of new paint. A dual-wheeled truck and the fifth wheel
were parked behind the detached garage. Betty explained, "Ralph is resting just now—best not disturb him. We retired from farming and moved
into town two years ago. Got a bargain on this house, so when we got
bored with sitting around we bought that RV and took off for Yellowstone
and Glacier a year ago this summer. We just loved it, but Ralph started
getting dizzy spells on the way home. I've got cataracts but managed to
drive us back. Now we've got medical problems and expenses and not
much hope to use the RV again. Hope someone else can use it."

Cal agreed that the vehicle was in good shape. Tires showed little
wear and the interior was immaculate. It had been used with care. "Now
I'm retired from teaching up at Jamestown, live over beyond Lynn. My
neighbor lady and I lost our houses in the big storm. Need some temporary, maybe permanent housing. I'm shopping on her behalf so can't
make a commitment now. We probably can't afford the tow vehicle, and I
have a truck as you see. She could probably come by to look at it in a few
days, next weekend at least. Can't expect you to hold it for us if you get a
good offer. Maybe you could give us a chance at it though, give us a call?
I could give you a check for consideration."

"Oh, no. Your telephone number and a handshake will do. I understand about the truck. Actually, farmers around here will have more use
for it than for the RV. And that storm didn't do much for our drought out
this way so a used truck will fit their pocketbook better than a brand new
one. We won't have a problem with it. Now, I think your lady would like
the RV, so give us a call for her to look at it. Then we'll go from there."

RV Purchase

Calvin called Kathy that evening. "You've been away for almost twelve hours; I'm missing you! Sheriff misses you. Hope we haven't done anything to annoy you, like keep you awake at night. But I'm not sorry; tell me you'll come see us again real soon!"

"Cal, you're sure? Thought you might need more time to recharge you batteries, or whatever. But then, I'd like to think that you will always be that way for me. Yes, I enjoyed the evening immensely. Give me an excuse to come again—pardon the pun!"

"No excuse needed; I'm always available for you. Actually, I've found a fifth-wheel like we were discussing; it's over at Masonville. You might like to take a look at it—give you a better idea what living in it might be like. That is, if you still have any interest in that sort of thing."

"It's a distant second. Do we look at it before or after? Oh, Cal, I'm being terribly silly, but I'm a new woman. And yes, I want to look at all options. I'll be down by lunchtime tomorrow; maybe we could give it a look-see in the afternoon?"

Betty Carlsud was eager to open the RV for Cal and Kathy—two o'clock would be fine. After introductions she explained, "Now working in a small space like this will be a change for you, Kathy. But I've managed nicely; everything is so convenient. Efficient kitchen facility including that big microwave. Toilet is small of course but everything works perfectly. There's air conditioning and satellite TV; just like downtown as they say. Now I'll leave you to talk it over by yourselves; I'll be in the house with Ralph."

"Kathy, what do you think?"

"I'm thinking of it as sort of an adventure. It would be mostly for weekends; I will still be at University in midweek. I'll give it a try. Can always resell it I suppose. But what's the price?"

"We'll have to dicker with the Carlsruds. Used ones on the Internet are listed at about half the price of new ones. This would be 'as is'—no warranty. Don't want to take too much advantage of Betty and Ralph though—they look to be hurting just a little. Still, they could think of the loss as the alternative to leasing one of these for their big vacation last year."

"And I hope you remember that our insurance companies are making us hurt just a little too!"

It turns out that Ralph was more than eager to unload what he considered the white elephant in the back yard. "I don't like to be reminded that I'm not in condition to play with that toy anymore. I think we're close enough on the dollars to shake hands on it. Oh, but you'll need that towing platform. We won't need it; throw it in as part of the deal. Write me a check and get that expensive monster out of my sight!"

Kathy responded, "Betty, the check is good; I'll phone the bank to make sure they don't go into shock about it. And Cal, suppose your young hulks can help us get it out to my homestead in the next day or so?"

"They can put current projects on hold. Plumbing and electric will be easy. We'll have you ready for open house before the weekend!"

Tom and Dick were eager on Friday morning for the challenge. Transfer of the towing platform was no problem; it was securely anchored to Calvin's flatbed. There was still enough space for Sheriff to roam about for short errands. The tow out to the farm was made without incident. Kathy's well and septic had survived the storm unscathed of course. The electric connections and metering were made to the power company's satisfaction. Plumbing was somewhat temporary; further precautions would be needed for cold weather. The guys insisted on escorting Kathy to town for the makings of a celebratory dinner. They would be satisfied with steaks and a beer; Kathy planned a more thorough stock of provisions. Calvin made a trip to storage for some of his mother's old furniture for an outdoor supplement of the dining facilities. He promised a temporary loan of his charcoal grill to cook the steaks.

The feast included corn on the cob, sweet potatoes, pole beans, and a fresh salad in addition to the steaks. Mrs. Smith's pumpkin pie and Starbucks coffee followed just at sunset. Then after the feast, Cal's helpers took off promptly for a hopeful double date in Masonville. Cal and Kathy moved inside with another glass of wine to check on local TV reception from Kathy's new habitat. As Cal seemed to be settling in for comfortable smooching, Kathy cautioned, "Cal, I want to experience my new home in private relaxation. You and Sheriff need to return home to your cave." Then as she observed Cal's obvious disappointment, she added, "I suppose we could test that new mattress for a few minutes of togetherness—do you think?"

It was about one hour later when affectionate goodbyes were exchanged outside the entrance of the RV. Cal approached his truck and called to Sheriff for a lift up. Silence. A careful search in the dusk found his faithful dog lying very quietly in the shadow of the RV. "Come on Sheriff; time to go home."

No response. "Sheriff, come!" A whimper. Cal approached his black friend, worried that Sheriff was sick or somehow trapped, unable to move. Sheriff continued to rest his muzzle on the ground, but looking up at Cal with pleading expression. Cal lifted the massive head and stared at his buddy. Then, moving parts of the heavy body, he found that Sheriff could cooperate without distress. Finally, with hand under the muzzle, he asked, "Do you want to stay with Kathy, is that it?" The tail wag was the positive response.

"OK, but we must ask her permission; now get your butt out here so I know you're all right! Come, sit!"

"Kathy, man's best friend has deserted me; he wants to be your guardian to protect you from things that go bump in the night. If you hear a ruckus, it will be Sheriff putting a coon up a tree."

Kathy approached, kneeling and ruffling Sheriff's ears. His response was a lively slurp in the vicinity of Kathy's ear. "Sheriff, your loving protection is gratefully accepted. And I expect your master may drop by occasionally, like for morning coffee?"

El Nino

Calvin's young helpers dropped by the next afternoon to see if there might be some chore to take care of before heading out to college later in the week. "Hey, guys. You're a few steps ahead of me. That's not to say everything is all organized and ship shape. But nothing that I can't take care of. Keeps me out of mischief, sort of. Now there is that big project of the solar cells, but have to wait for some of the parts that you ordered for me. And no need to do that before the weather gets a mite warmer. But say now! There's some left over Mrs. Smiths Apple Pie in the kitchen. Some Haagen Dazs to go with it. Don't suppose you are hungry enough to manage that though."

Tom answered, "I'll force myself—and work on Dick's as well if he's on a diet."

"Keep your distance; you might just lose a few fingers."

Conversation was on hold while Calvin divided up the pie and added ice cream. And there was nothing to be said until the last of the crusts were demolished. Then Tom adopted a relatively serious expression and posed a question for Cal. "What's with this El Nino thing that I hear about on the Science channel. I surfed into the end of a program—didn't understand the details. It's a climate thing, right?"

Calvin replied, "Very much so. But I don't know any more than you'll find on TV or maybe some magazine report. It's had lots of publicity lately, because we are just beginning to understand it. I don't recall that it was mentioned in any of my formal university courses—some years ago, sorry to say. Probably described in the most recent textbooks of course."

"Tell you what I know of it. The name originates with the behavior of ocean currents off the coast of Peru and Ecuador. Offshore fishing is the livelihood of folks there. Fishing is good when upwelling cold water brings food to the surface for the fish. But occasionally, every few years about Christmas time, that fails and fishermen suffer. The literal translation el nino means child; when it is in caps, El Nino, it means the Christ Child—the seasonal connection. And with the warm water there is a shift in the atmospheric circulation. The winds were east to west from high pressure over the cold water off the Peru coast to low pressure over the warm water in the western Pacific. (The Coriolis Effect is absent at the equator of course.) Winds are reversed in the El Nino. Now this is a major change in the worldwide circulation and changes the weather everywhere. Floods in California and droughts in Africa. And it's not a new effect; history suggests that it was happening for thousands of years. Droughts and floods tended to occur with a period of roughly five years; it became known as the Southern Oscillation—now the acronym, ENSO."

"So, what causes it? Must be the ocean currents, or by the atmosphere, or is it a solar effect?"

"Well, these things are all coupled together. Obviously, changes in the sea surface temperature will change evaporation rates and the atmosphere will soon have different precipitation and wind patterns. Then the ocean currents will begin to respond to the wind shift. It's not a steady situation; not surprising that it runs in a cycle."

"And there's a solar cycle too, isn't there?"

"True. That has a period of about 11 years, roughly twice that of the El Nino. And there are other atmospheric periodicities—something called the quasi-biennial oscillation out in the Pacific. All these things have to go into the climate models to understand the coupling, and to predict the future."

"And I suppose global warming has to be considered."

"Correct. And this is a new effect. We won't find this in the history of recent civilizations. We're flying by dead reckoning on this one. If we put garbage in, we'll get garbage out!"

"You say, 'we'! I suppose there is a government lab someplace that is doing this?"

"Government support, yes. At Princeton and the UK, for instance. But, we're not excused. Remember, we are part of the problem—putting those greenhouse gases in the atmosphere!"

"Gosh! Maybe I should ride my bike up to University instead of the old Buick!"

Jefferson Wonders

It was a few days into the new month when Cal stopped by the bank to move some money from savings to checking; a delicate but necessary operation. About to depart, he was approached by the doubtful environmentalist, Basil Jefferson. "Say, Cal! A question for you. My informants from talk radio continue to make fun of the global warming topic. Like, it's just a natural trend, nothing new. Mother Nature is too complex and powerful for a few humans to upset things. And why now? They suggest it's all politics."

"Well, there's a book, *The End of Nature* that they haven't read. The idea is that the human populations have increased, and we have used a heavier hand on our surroundings than most other creatures. The rules have changed. Now we have to be considered as part of Nature. As for it's being a natural trend, there has to be a valid physical reason. I rather doubt that your talk show authorities can explain that for you."

"Well still—so long as we don't upset the economy, we can adapt. Buy bigger air conditioners, import our wine from Chile, take vacations to Alaska, whatever. Just keep the machinery running and the cash flowing! We've got the resources; fix whatever is needed."

"Jeff, you need to understand the time scale that's involved. If CO_2 is warming the planet, we can't just shut off the effects. It hangs around in the environment for something like a hundred years. It's like a big lake behind a dam. Maybe you don't ever expect the dam to fail, but if it does, the flood will wash your house away. Now you understand insurance. I'll bet you would have flood insurance on that house! Maybe we ought to slow down on burning fossil fuel; sacrifice a little, just in case."

"I'll have to think about that. If my investments show signs that the dam is leaking, I may hedge a little on wind power."

"Basil, the melting glaciers are increasing sea level just a bit. It's later than you think!"

Calvin's thought that evening, "So another topic for Chris. Hope Jeff doesn't wonder how such a prompt answer comes about!"

It's Later Than You Think!

Go ye forth and multiply. And we did! We conquered the other animal kingdoms and the myriad forces of Nature. Now we find ourselves to be the potential victims of our success.

Human Populations

The increasing human population on earth has become an important influence on the natural environment. Other life forms have also changed the environment. A long time ago, the bacteria and plants utilized the process of photosynthesis to begin adding oxygen to the atmosphere. Oxygen concentrations increased rapidly from negligible amounts to present day 21%. This allowed the evolution of aerobic life forms leading to present-day animals and humans. We naturally think of this as a positive development; it even culminates in the production of the stratospheric ozone layer that protects us from the damaging solar ultraviolet.

We have recently escaped the near disastrous destruction of this protective ozone layer caused by inadvertently adding rapidly increasing amounts of man-made CFCs to the air. These molecules were extremely useful in modern life for refrigeration and other industrial activities. It apparently had no deleterious effect on the near environment. Unfortunately, upon diffusing into the stratosphere these compounds were broken apart in a natural absorption of solar ultraviolet, releasing the chlorine free-radicals that destroy the protective ozone layer. Human societies have subsequently learned to cooperate in reducing this threat to our environment.

In the above examples we have seen rapid increases of concentration of certain atmospheric constituents, one was good, the other bad. We now see a similar rapid increase of CO_2. The coal and oil stored in the earth are being burned at an increasing rate to fuel industrial activity. It is the reverse of the very long process of storage of atmospheric CO_2 by photosynthesis in the distant past. This increase in CO_2 is well below the threshold of direct harm to humans and it is in fact a stimulus to plant growth. However, as in the case of CFCs, there is a natural process that poses an indirect threat to earth's climate. The greenhouse effect of this molecule is very effective in reducing the escape of infrared radiation, permitting an increase in global temperature that threatens to yield disagreeable climate changes.

The increase in human populations is driving a demand for more and more energy for transport and industry. There has been a mixed response to dealing with this human-induced threat. The production of energy by oxidation of fossil fuels has been considered the essential driving force for progress in the industrial era. Industrial nations have been reluctant to limit these methods of power production. Some countries have begun modest efforts to reduce CO_2 emissions. But, the greatest offender has refused to alter its industrial activity. And other large developing nations continue to follow this example. This is not a good plan for the future!

Yours for analytic thought,

Chris

Town Meeting

Cal had stopped in at Ben's Café for midmorning coffee after shopping at the hardware store for a supply of finishing nails. The place was a beehive of activity. Casual remarks about the weather had grown into a discussion about the reality of global warming. Tables were being moved to the center of the room, lending the importance of direct eye contact to the proceedings. Ben was refilling coffee cups as Russell Harding was pooh-poohing the dry weather as part of global climate change. "The rain is always variable. Now I'll admit we had a dust bowl a while back, but we survived and things are better than ever. And West Virginia is having flash floods—nothing global about it!"

Dave Crause interrupted, "Here's Cal. Give us an update on this climate thing!"

Cal replied cautiously, "Sounds like you've been exposed to some climate stories—the TV and newspapers have been giving it more attention. Go ahead—what are your reactions? Ignoring it won't help—some discussion would be good."

Banker Jefferson made a casual gesture, "Oh, sure. We have our ups and downs. But we are the superpower. Put our economy to work and the climate will come to heel. No need to get excited."

But Mayor Clarkson was excited. He stood and sounded off with a typically long speech. "So, we live in the land of a superpower. Think we deserve it? Being a superpower is easy if you live in a land that has abundant natural resources for the taking. Then when the necessities of life have been satisfied, you have the means to form multinational businesses to buy additional goodies from far countries. And if another ambitious nation threatens to move in on our good fortune, we become determined to protect our way of life. We allocate a significant part of their wealth to the military. It's the combination of economic and military wealth that makes us a superpower."

The mayor held up a cautionary hand. "But the drive to accumulate an abundance of household and military toys requires energy. Not the kind that you need to become a captain of industry, but real energy. The kind that is relatively cheap if you have ample resources of fossil fuel. And if you run a little short of one, a little military or economic arm-twisting usually solves the problem."

Clarkson was just approaching the point of his speech. "Now I've been reading that our scientists' awareness of the greenhouse effect of various atmospheric gases prompted concerns already in the last century of the potential danger of increased CO_2 from burning of fossil fuel. And I understand that the patient long term monitoring measurements at Mauna Loa has verified the projected CO_2 increase and fossil fuel link. But Chicken Little's warning that 'the sky is falling' doesn't seem to have generated much concern from economic interests."

The mayor concluded, "So we have muddled our way deep into a swamp of problems. The planet is past the point of no return in putting greenhouse gases in the atmosphere. And if the CO_2 is really the problem, the Chinese are the ones who are making it worse. Right? Now, I wonder why it's always the other guy's fault. But who is going to fix it? Let the government do it! Well, we are the superpower; maybe we could work on it sometime." Turning to Basil Jefferson, "Sorry Jeff. I've been doing some thinking about what I hear on the news. Doing some analytic thought like that fellow Chris has been suggesting on the Internet."

George Stanhope of East Kansas Power sounded off, "Oh come on Mayor. We can't just leave the coal in the ground and shut off the electricity. These problems always go away after a bit. Be patient."

Dave Crause from the pharmacy asked for attention. "We hear things about climate change from remote places. But, the climate includes ocean circulations and the evaporation and condensation of water vapor. Each of these is responsible for much of the earth's climate and daily weather. We now seem to be witnessing the unpleasant surprise of a drought just west of here and an increase in violent weather. We tend to ignore the accumulated scientific data, but a series of category five hurricanes got your attention, didn't they!"

Russell Harding insisted, "Those government scientists haven't been doing their job. When those storms started up out there in the Atlantic they should have done something to steer them away from the U.S."

Cal interrupted, "Sorry. Once you let the atmosphere get charged up with water vapor, there's nothing humans can do to control it. And government scientists don't seem to have very good press. It's true that scientific studies about the environment plod slowly along with limited funding and occasional fundamental revelations. And such discoveries usually take a long time to show up in the profit column of the economy. But sometimes the scientists take an active role in warning about the environment. May lose them grant support of course, but with a little luck people pay attention. For example, the scientists' concern about damage from nitrogen compounds to stratospheric ozone found unusual support in doubts about economic viability of super sonic jets. A similar theoretical concern about chlorine compounds received attention only after a surprising discovery of major ozone damage in polar regions. That prompted a gratifying international response to discontinue manufacture of the offending compound. Unfortunately, that mistake cannot be completely corrected; we cannot promptly return to square one. A reservoir of long lifetime chlorine compounds now exists that will allow only a slow return to normal ozone in about 50 years. In the meantime, we get to use sunscreen or stay out of the midday sun. But global warming could be a much nastier problem!"

Farmer Jones had his say, "My 6 AM Farm News program on the radio has been telling about noisy temperature records, melting glaciers, disappearing Arctic ice, and warmer oceans. That sounds like global warming to me. But, these things are so remote; a shrug of shoulders and, 'just natural variations' don't get me much stirred up. But when I listen to those science programs on TV, I hear that difficult and expensive scientific measurements continue to give solid results that can't be ignored. And when they dig out the historical scientific data that describe the earth orbit and solar radiation they see what caused those earlier natural varia tions. Now I hear that my tax money has been used for increasingly elaborate and sensitive observations, and that these have begun to firm up the global warming threat and that we are causing it. But still, the powers that be continue to dominate our government policies without paying attention to the science. And digging up the hills for coal, and drilling holes in the ground for oil and gas is a frantic way to maintain the industry profits and the activities of the military. Looks to me like we've stopped being a leader in caring for the land and the air. Some of the smaller countries seem to be thinking about the science though, and trying to do the right things. I'm beginning to feel a little guilty about us."

Jefferson replied, "Those little countries don't count for much. We are the richest, most technology advanced. And the developing nations pay attention to us."

Dave Crause answered, "Well, the majority of smaller nations have already made slight course corrections in energy use, but we have continued merrily on our way deeper into the swamp. Yes, a world leader inspires others to follow. If the inhabitants of some Asian country are above starvation level, they begin to envy the jet aircraft, automobiles, televisions, and mansions of the superpower. If the energy resources are somehow available, they follow the example of the successful superpower. They eventually compete for the energy. And they compound all the mistakes made along the way!"

Cal nodded agreement, then cautioned, "Eventually, we must make major sacrifices to conserve and develop alternative energy sources. Even then, the long lifetime of CO_2 in the environment works to blunt any early future relief. We must live with our mistake in an altered environment for the foreseeable future."

"Thanks Cal." John Axelson, the auto mechanic from the edge of town wiped a greasy hand on his shirt and pointed in Cal's direction. "I hear you loud and clear. So, there were politicians in Washington who got us into this mess. Well, a bunch of us were in that mob of sheep who just followed right along. And we contributed to the problem at least as much as the city folks, or the Chinese who are burning all our gas. If we had listened to our scientists and given some, what that Chris fellow calls analytic thought, to the situation, politics and cheap oil wouldn't have seemed so important. We squandered our opportunity for being a leader, and the developing nations have followed in our wake to near catastrophe." He pushed back his chair and stood. "Seems to me that's the bottom line. I've spoke my piece; now have to get back to work on fixing Jeff's Hummer. Says his gas mileage is so low there must be a leak someplace!"

The Blue Planet Radio announcement of the upgrade of a late season Atlantic hurricane suggested an opportunity to prod the reluctant non-thinking public. Hopefully, by this time they may have accepted this climate change threat.

It's Later Than You Think!

So, we're already committed to global warming. No sweat! We can adapt—some good, some bad. Oh sure, some natives in the Arctic or the sub-Sahara will have to lump it or just disappear. But those of us in advanced civilizations will manage just fine.

Adaption

You're willing to accept this little handicap of a few degrees of global warming. And the superior genes of your ancestors on the Mayflower will allow you to adapt in good fashion. We've had dry spells on the western plains before. You just sink the wells into the Ogallala a little deeper and buy bigger pumps. And by switching political parties you can get laws to stop the Texans from emptying the aquifer! Or maybe the government will build a pipeline to bring water from the Mississippi.

Or if things get really bad, you can emigrate to Canada. With global warming, temperatures in Ontario and British Columbia should be warm enough to grow wheat. Have to cut some of those trees of course. Or if the Alaska permafrost melts you can lease some of that Eskimo land and grow more wheat. And if the polar bears can't find seals for food out on the ice, 'cause it's melted, you could probably do some trophy hunting on the side. Actually, the natives will probably welcome you; those bears will be desperate, probably settle for Eskimos if they can't get seal meat.

Of course, we expect our good neighbors to the north to welcome us with open arms. Surely, they would welcome some new blood from the U.S. Better than polluting the gene pool with the Chinese! But did I hear something about a wall?

You've already given careful thought to those possibilities? I should have expected that! If one makes a mistake, the same logic tells you to stay the course!

Oh, I know. The alternative isn't good. One might have to settle for subsistence farming and sell the John Deere and the Cadillac. Of course, everything is relative. Those Africans in the sub-Sahara, who can no longer grow their own food, might have survived on a few pounds of rice from the UN. But we're busy trying to preserve our economy; so, they'll starve.

Yours for analytic thought,
Chris

RV Escape

Spring arrived with blustery winds and brief showers. Cal began to browse the Internet for evidence of northerly circulation out of the Gulf, or cumulous activity in the eastern Atlantic. An intrusion of a moist tropical air mass to collide with cold Polar air, or the appearance of the first tropical storm would be the sign for preparations to migrate with the RV out of tornado alley and away from the tracks of dissipating hurricanes. Fortunately, tornado alley and early spring storms seemed to have moved eastward, due to the continuing deep penetration of the polar jet stream into the western Great Plains. But in mid-April Cal became anxious to move out of the path of potentially violent storms.

"Kathy, it's time we examined our maps for camping spots someplace in the West. We need to move your RV out of storm danger, but just putting it in storage is a waste. If we start someplace in New Mexico or Arizona we can move north to Colorado with the sun. Then, providing things have quieted down out here on the plains, we can come back in time for you to prepare for University classes."

They discovered that arrival at Bosque del Apache near Socorro, New Mexico would position them just prior to the departure of the sandhill cranes to northern nesting areas. Then a visit to Canyon de Chelly and the Grand Canyon would guarantee sunshine at these famous sites before being overrun by tourists. Finally, a move into the low elevation parks of Colorado would permit an early return home, or delay into the high elevations near wilderness areas if appropriate.

They embarked on their journey. A careful avoidance of commercial RV parks encouraged access to natural areas, albeit in the presence of like-minded RV owners. Being aware of fuel uncertainties and cost, they started with both truck tanks filled to capacity. But as the travel season progressed, gas supplies and costs became more of a problem. Still, the precautions with the RV appeared to have been a wise move. In early May, they received a cell phone message from Dick describing the arrival of an early season tornado. "Our spring weather had been pleasant with warm temperatures and a few showers. Then a fast moving cold front came through and we had ourselves a late afternoon twister. No problem at University; it's good that we left Kathy's vehicle up there. Here, it touched down just at the edge of town, then took dead aim at your place, Cal. Looks like you've got bits and pieces of somebody else's furniture scattered about in front of the cave. The cave was untouched of course. That old windmill is lying on it's side; 'spect we can find a replacement from a farm with a dry well. It's good you were off in the mountains with the truck and RV. But then the weather here turned hot and dry. We've had triple digit temperatures all over the plains. Folks here do the standard thing-chores in early morning, then sit under a shade tree with tea and lemonade and hope for a breeze. KC is in trouble though with power outages because of the air conditioning demand. Movie theatres with emergency power are publicizing free old movies and bottled water in the afternoons. Business and shopping has come to a standstill. No need to hurry home; enjoy your trip!"

It was in Colorado that their travels hit a snag. The meager winter snows had melted in early April; by mid-June the Continental Divide had no snow cover. The rosy finches were discouraged from nesting and the pica had run out of food. But access to the high country was available to the RV and the dry warm camping was pleasant. Unfortunately, the Forest Service campgrounds had no water; the wells were dry. The foothill communities had grown rapidly in recent years and each household well had depleted the diminishing supply of ground water. Shallow wells were dry and demand for water from deep wells was great. Cal and Kathy found that re-supply of RV water was even more costly than purchase of fuel!

The traditional monsoon did not develop on the east slope in early July. Wildfires began with lightning strikes in southern Colorado. The ponderosa and lodgepole forests had become especially vulnerable because of insect infestations encouraged by a series of mild winters. Within weeks, the foothills of the entire east slope of the Rockies were in flames. The air was filled with smoke. Cal especially, found breathing to be difficult; Sheriff was worried and restless. Slurry tankers were only occasionally available due to fuel shortages. Other equipment and manpower shortages endangered entire mountain communities. The uncontrolled crown fires continued to move up the mountains. Finally, complete evacuation of the high country was ordered. Convoys of residents and backcountry campers were escorted to the flatlands through heavy smoke by the U.S. Forest Service, aided by the National Guard. Cal and Kathy found themselves temporarily marooned in Pueblo with insufficient fuel to make the trip across eastern Colorado and Kansas. Emergency supplies arrived ten days later and they proceeded carefully homeward through the dust-laden air of the Great Plains.

****

On return to the 'cave', a haphazard search of AM frequencies finally reached Blue Planet Radio. The conclusion of instructions for replanting overcrowded cottonwoods was followed by an announcement of a special interview with Climate Consultant Dr. Andrew Fitzwilliams. "We've asked Dr. Fitzwilliams to respond to the published scientific report of increased melting of the Greenland ice cap. It is generally understood that this will result in a rising sea level. Beach erosion will increase; there may be other effects that Andrew will address."

"Thank you again Marny. I'm honored to be here. I'll try to be helpful with this complex subject. We've always expected rising sea levels from thermal expansion of the ocean water and melting of the glaciers. Rates have been something like a few millimeters per year—not an immediate problem. But we have been surprised to find that these effects are accelerating; now the Greenland ice cap is kicking in. The evidence on Greenland is solid. Our satellite instrumentation is good, much better than anything we've had previously. The increasing motion of the Greenland glaciers is an unpleasant surprise. A cushion of melt water or intrusion of warm seawater may be responsible. The time scale of increasing sea level may be much shorter that previously believed. Residents of Pacific atolls are becoming very worried. Still, the sea level rise will only slowly become apparent on our shores; no need to abandon your beach house just yet.

One further clarification though. The notion has surfaced that the rising sea level will trigger major geologic events. In particular, there is now a fear among much of the public in Southern California that part of the state will be separated from the mainland at the San Andreas fault because of the shifting weight of sea water. There is now a panic migration to inland areas. Consequently, we have political pressure on the BLM to open more Nevada land to development. Is this the intelligent thing to do?

The above rumor is not founded on any geologic evidence. While many uncertainties persist abut earthquakes, I would not expect that a few extra inches of seawater at the coastline could produce this result. Of course it is happening all around the Pacific rim; perhaps some real problems in the distant future with all the melt water from Greenland and Antarctica. And I wonder what will happen to those polar land masses when the weight of all that ice is removed! I am not a geologist however. I call on the geology scientific community to assess the situation and publicize their conclusions. The news media need to respond responsibly. The public needs to pay attention. Any needless abandonment of properties and disruption of other environments would be a poor use of our resources at this time.

I thank you for your attention. Marny, back to you."

Calvin was disturbed that he had not anticipated the speed of the developing consequences of climate change. The slight increases in global temperature had been predicted; the extensive wildfires and dust storms seemed to be a precipitous response though. The efforts of some segments of society to adapt were even less predictable. The fuel shortages were inconvenient, but at least those effects on greenhouse emissions were in the right direction. More constructive efforts in energy production were going to be necessary though, and the sooner the better!

It's Later Than You Think!

Scary Things

Those category five hurricanes and explosive forest wildfires are inconvenient. I know, we'll muddle along and get things under control eventually. Actually, things seem to be moving along about as predicted by the global computer models. But perhaps I should remind you of the lesson to be learned from our little scare with stratospheric ozone. The polar ozone holes were a surprise; they were beyond the predictions of the theoretical models. And even though our international response to cease production of CFCs was prompt, we are left with a 5% reduction in global ozone that will require decades for nature to repair.

But with climate change, the situation doesn't seem urgent; especially since those little disasters of hurricanes, forest fires, famine, and disappearing islands are happening in someone else's backyard. Of course, when our air conditioning bill gets outrageous and tomatoes become unavailable in the supermarket, it becomes expedient to react. So, it is possible for most of us to adapt to change in a time scale of 10 years or so. But I need to remind you that there are some scary things out there beyond that time!

The surprises aren't so much what, but how soon. At some point, rising temperatures are expected to trigger the lubrication of the Greenland and Antarctic ice caps resulting in a rather sudden massive dump of fresh water into the oceans. Sea levels would rise abruptly by a matter of feet—or meters, submerging Pacific islands and displacing coastal populations worldwide. That additional fresh water would likely stall the return flow of the Gulf Stream. The resultant cooling of the European Climate would be most unpleasant. On the other hand, the Arctic sea ice is melting; scientists see no way to avoid ice-free summers sometime soon. Already, the polar bears are in distress because they can't get to seal meat because of open water. Also, at some tipping point of the rising temperature, the melting of the tundra permafrost in Alaska can be expected to release the CO_2 and CH_4 sequestered in the organic soil. Finally, at sufficiently elevated ocean temperatures, the oceans might begin to release the immense amount of CO_2 and CH_4 stored therein. The tipping points of such elevated temperatures are generally believed to be somewhere in the distant future, like centuries. Still, there are some observations that suggest an early surprise! Such positive feedbacks to the greenhouse effect would be disastrous—much too fast for the polar bears, or many of the other creatures— like us—to adapt.

Now, such predictions are obviously designed to get your attention and frighten you into reduced procrastination. But like the near disaster with CFCs, CO_2 has a very long environmental lifetime. We are already committed to the global warming trend. Fortunately, we think we have a bit of slack before surprise time. We are hopeful that we have 10 years or so to correct our mistakes and invoke the 'precautionary principle'. Perhaps we can redesign our energy production to survive without pushing the greenhouse effect over those tipping points.

Yours for analytic thought,

Chris

Chapter Six

WINTER

Season Of Change

October Indian summer had been very pleasant. Cloudless skies and warm temperatures were enjoyable. But Cal had noticed a gradual influx of cirrus from the southwest and a polar air mass was on the way. Early winter weather might well be violent. Moving the RV out of harms way was getting to be a nuisance though. And the wildfires in the mountain forests were not inviting. Several steel cables over the RV roof and anchored securely in the ground would have to suffice. A direct hit by a tornado would still be a disaster, but perhaps the structure would survive blizzard conditions.

And there were a few late season tornados, mostly to the east. The small Arkansas town of Cotton Corners was obliterated. Cal and Kathy were snuggled securely in the 'cave' during the gusty late season thunderstorms. The RV suffered superficial damage; it leaned a bit to the east, but the winch on the truck would fix that. Folks in the FEMA mobile home parks were not so lucky. Their possessions were scattered across the county—again.

****

It was a late fall evening when Cal tuned in to the Blue Planet Radio. "That's it for the stats on last weeks weather. Now, Dr. Fitzwilliams is here to update us on this past summers unusual weather and his thoughts of the future. Andrew, is it really true that we are committed to nasty weather caused by global warming?"

"It is an honor to be here. However, I hope you will not feel obliged to shoot the messenger for this update on climate change—we are experiencing continuing evidence of global warming. I wish you could arrange for more pleasant weather in your introductions. But here goes: Sea Surface Temperatures in the eastern Pacific are reaching record high levels. The El Nino is in full swing a bit ahead of schedule! Fortunately, the resultant westerly winds in the tropics should inhibit the formation of large Atlantic hurricanes next summer. But they have brought increased moisture and torrential rains to California and the Oregon coast and Mexico. The battle lines of strong polar air masses and this tropical air are drawn across the eastern Great Plains. So we continue to have early fall violent storms. Winter in the eastern U.S. should be relatively mild, but with increased precipitation, especially in the southeast. Then the warmer springtime temperatures and high humidity will set the potential for a super active tornado season. Sorry about that!"

"Now some general comments about this global warming phenomena. The deep penetration of dry continental polar air into the Great Plains moved us quickly into a drought situation. The effects have been particularly apparent in the foothills of the Rockies and in the High Plains. Wildfires and dust storms have been prevalent. The Arctic has warmed however and we may get some respite from that cold dry wind. Precipitation is doubtful though. And replenishment of the Ogallala aquifer by precipitation on the east slope of the Rockies is essentially nil. These polar air masses have also forced the hurricane tracks eastward. The Gulf shores of Texas are off the hook for hurricane damage but the Florida peninsula is being hit hard. The moist tropical air is moving into the Appalachians and up the east coast with extensive rainfall with floods and mudslides. As winter approaches, these storms will become larger and more violent. Then moving into the Atlantic, the Nor'easters will repeat last winter's snows and eastern shore beach erosion."

"Now some global warming effects were expected. The greenhouse effect is responding slowly to the increase of CO_2 and other products of industrial activity. Of course, at some point the consequences can be severe, but the tipping point is uncertain. Surprises could happen; the Greenland ice could move abruptly and the fresh melt water from Greenland would not be good for the Gulf Stream. The return thermohaline circulation to the south on the ocean bottom needs dense water, like cold saltwater. It may already have slowed; England is facing an early winter with snow and frigid temperatures. While the increased snow cover may increase the earth's albedo and counter the complex results of climate change, the reaction may be too slow to relieve the tragic consequences of frigid weather and heavy snows in England and Europe."

"If the Gulf Stream is beginning to fail, this means more pressure on the atmospheric circulation to transport heat from the tropics. A major part of that involves the movement and condensation of water vapor; I'm talking about weather folks; Nor'easters in winter and tornado alley in spring and summer. And the sea surface temperatures have risen and the layer of warm water is deeper. That deep reservoir of energy will be only slightly diminished by a storm's traversal. Hurricanes will be larger and more intense—category five or whatever."

"Our weather in the Great Plains is highly variable of course. However, there is little doubt that this general trend is a product of global warming. Computer predictions of local conditions are uncertain, but seeing is believing. I'm sorry to be pessimistic, but there is little basis for expecting matters to return to the climate of a few years ago. As you know, the environmental lifetime of carbon dioxide is something like a century. Our efforts at mitigation of the pollution have been little more than a token. We must begin to concentrate on adaptation."

****

In November it began to rain in Texas and states eastward. In a few weeks the ground was saturated. Rivers hovered near flood stage. Cal and Kathy lived near the western edge of the deluge, but the Raccoon lapped at their doorstep frequently. January weather was normally cold and dry. This year it was wet with freezing rain instead of dry snow. Kathy was frequently marooned at University due to treacherous roads. Food supplies were erratic. The Masonville supermarket maintained a reliable supply of canned goods and staples but fresh produce was hopeless. The supply of strawberries and vegetables from Mexico was disrupted by floods. Substitutes from Chile were erratic and expensive. Florida crops were damaged by fungous encouraged by exceptional humidity and high temperatures. But the drought in the western high plains continued. Springtime vegetables from irrigated farms near the Colorado foothills were unlikely. Calvin began to think of a springtime garden.

Celebration

Kathy and Cal tended to ignore the weather in their developing relationship. The atmosphere in the cave was cozy and unchanging. Kathy's RV continued to furnish its modern amenities in spite of the cold dusty wind outside. But then, as winter university break approached, Kathy became restless and searched for warm vacation spots in the sun. It was late one Sunday evening when she remarked, "Cal, these evenings of love with you are very nice, but I wonder if we can't find a balmy beach someplace to enjoy each other."

"With you, I could probably tolerate that."

"I'm beginning to wonder if you have a one track mind—not that I object of course! Anyway, there is a couple who attend my adult education music appreciation class; they have been raving about the beauty and culture of New Zealand. And I've been checking on tour arrangements. Travel is expensive, but accommodations, especially bed and breakfasts are reasonable. And Tahiti is on the way! Now if you could commit to be faithful enough to not run off with some Maori lady or Tahitian beach girl, I would like to blow some of my remaining stocks and bonds on about three weeks of ecstasy. Now I know you plan to live forever, but I'm not getting any younger. I'd like to enjoy this before I have to hobble about on a cane. How about it?"

"You come up with the most delightful ideas. Living with that suggestion just puts our life up on another plane. It's true that I have been living in anticipation of every weekend. And I've been able to nicely ignore our dreary winter. But a few weeks of summer would be just lovely about now." Then hoisting himself up on one elbow to stare down into Kathy's eyes, he continued, "I wonder, could we think of this as a celebration of the beginning of the rest of our lives—together? Kathy, I'm sure you know that I'm very much in love with you. I'm not enough of an optimist though to think our days ahead will always be sunshine on the beach. Still, I would like to think that our relationship could endure anything that comes our way."

Kathy was smiling. "Yes, Cal. I've thought about this very carefully, as you know. And I think we have already made that commitment. Now let's live a little!"

Of course, climate in New Zealand was also changing. But the extensive southern oceans slowed the rate of change. The ninety-mile beach of the North Island was slightly narrower and the dunes were moving inland. Showers in the Bay of Islands were more frequent and blustery. The stormy season of the roaring forties around Stewart Island became noticeably longer. But the temperate rain forests of the South Island west coast continued to prosper; the Southern Alps accumulated more and more snow. Christchurch and the Canterbury Plains continued sunny and pleasant. Cal suggested base operations with bed and breakfasts in that area, with excursions into the scenic areas of the South Island as weather permitted. "Now the rest of the world lives off credit cards like there is no tomorrow. We'll have lots of company when we return; let's just do it!"

The initial reception down under was a bit cool—something to do with ignoring the precautions taken by the rest of the world about climate change. Still, the dollars were reluctantly accepted. But then the sociability of New Zealand bed and breakfasts brought the true character of visitors into play and Kathy and Cal became respected world citizens and enjoyed the genuine friendship of their hosts.

The stopover in Tahiti for the return flights was relatively brief. But the waters were warm, the snorkeling views were gorgeous, and the days were long and sunny. The rain showers were refreshing and the tropical hillsides were spectacular. Truly, a climax for a wintertime vacation.

Winter Weather

The new semester at University required Kathy's renewed midweek attentions. Cal continued work on his cave, with assistance from Tom and Dick on weekends. His bedroom and study now had carpet; the kitchen had vinyl tile, and the east room enjoyed some varied width flooring resurrected from Kathy's damaged house. In spite of low sun and cold dusty winds, the boulder greenhouse maintained temperatures well above the cave's underground background. The fireplace was good for close relaxation with coffee and a book. Solar panels were planned for the distant future. Kathy's RV was abandoned for long loving weekends in Calvin's 'cave'.

The winter weather on the high plains was distressingly dry. The cold northerly winds recycled every three to four days with brief periods of faint sunlight through the dusty atmosphere. Pastures and fields remained brown and dry with shallow layers of drifting soil. The collision of the periodic outbreaks of polar air occurred with warm moist air in the eastern Gulf, initiating extratropical storms that moved up the eastern seaboard and into the North Atlantic. The long fetch over the Atlantic of the northeast winds battered the coastline from Massachusetts to Florida. Beach erosion was continuous with loss of thousands of beach homes. The storms became intense over the British Isles with heavy rain, then continual snow over Scandinavia. On the U.S. east coast, a continuous series of early winter nor'easters kept the snowplows moving on all the Interstates from New Jersey to Massachusetts. Residential streets became the sole responsibility of householders with snow shovels. But in late February, the snowfall made an unprecedented transition to rain. The winter snow became slush and the combined slush and rain spawned disastrous floods. Travel and commerce came to a grinding halt for the next two months.

Calvin's neighbors in Lynn and on the surrounding farms had endured many such winters. The outlook for springtime relief was not good however; climate experts were predicting violent early spring storms to the east, but continued drought over the Great Plains. Banks were reticent to advance loans for springtime planting; cash flow was restricted to bare necessities. Russell Harding spent many hours washing the dust off the tractors on display; he had abandoned his late winter vacation to Hawaii.

Basil Jefferson, from the bank, had made the usual visit with wife Miriam to their Atlantic beachfront condo in south Florida. Sun time on the beach was erratic; cold continental air masses arrived frequently with temperatures inland hovering in the high 50s. Temperatures moderated as the winds switched to northeast; the winds were steady and strong however; the beaches were eroding and the blowing sand was unpleasant. Basil and Miriam remembered enjoying videos of the coral reefs in the Keys; perhaps it would be good to scout the possibilities of financial involvement in fishing and glass-bottomed boats for retirement years. There was little business activity, though. Basil's inquiries led to the explanation, "The coral is dying from general pollution and climate change; summer temperatures have been too high. The fish have disappeared; nothing interesting out there any more. And they tell me that with the increasing CO_2 sometime in the distant future the ocean is going to be so acid that all the little ocean creatures will disappear. It's a damn shame!"

The Jeffersons signed on for a Caribbean cruise, in hopes of more pleasant weather. Seas were rough! Before returning to Kansas, Basil arranged for a realtor to put the condo on the market; property in South Florida was moving slowly however.

Society Breakdown

It was early March when they moved the RV down to Cal's place. The decision was made not without provocation. In the normal farming community operation, the social and safety considerations of remote residents were monitored as a matter of general practice. But the drought conditions prompted some of Kathy's neighbors to move away. The stability and morality of the Great Plains population was disrupted. It was a conversation overheard in the Cattlemen's Bar that caused Cal's young buddies to take action, and prompted the ultimate shift of the RV.

Al Scofield was the Cattlemen's bartender. Football games were finished. Basketball later in the evening, only a few customers this late Sunday afternoon. "Hey Dick. Where is your sidekick Tom? Isn't it his turn to buy the beer?"

"He's out shopping for gas so we can get back to campus. He'll be along. How's business?"

"Spotty. Some of my regular customers have decided it is cheaper to stay on the wagon. But I did have a couple of young punks in earlier; more cash than manners. Think they are from west Texas. Nasty weather out there probably jarred them loose. I get the impression they are staying with old John Baker; probably his nephews or something. Think maybe you should be interested."

"Oh? How so?"

"Well, they got a little talkative after their second beer, loud enough that I could hear real well. They were cooking up some mischief about a lady living alone near here. They heard that she stays in an RV on weekends; has a dog for company. They figured on managing the dog with a ball bat, then having some roughhouse fun with the woman. Even flipped to see who went first. It's for next Friday night."

Calvin was informed straightaway. "I understand. Not good!" Of course Sheriff was clearly a comfort to the security of his adopted mistress. He could be formidable protection. Any intruder, man or beast, would be detected and warned by a deep menacing growl. Overt physical attack on his mistress would be met with 150 pounds of solid moving muscle and jaws of crunching teeth. He would not indiscriminately attack a trespasser however. On the other hand, his heredity gave him the courage and determination to refuse to retreat. Unfortunately, he would be ill equipped to deal with a cunning and vicious aggressor equipped with the means to inflict damage from a distance. Such an attack would have to be met in kind. But a senior citizen armed with a pitchfork wouldn't slow them down much. "I'll have to see to some precautions."

Dick explained, "Cal, we have a plan. Your part is to be sure to entertain Kathy at your place Friday night. I suspect you can think of something that Kathy will find attractive; whatever it takes! There will be a committee of your friends to welcome any visitors at the RV. And don't hesitate to contact me by cell phone if you suspect a problem; remember, these are not nice people!"

It was Dick who arranged for four tickets to the Jayhawks-Hawkeyes battle in two weeks time. Then to a classmate, "Ted, I have a proposition for you and one of your karate friends. I'll bet you could learn a thing or two from watching some of the tricks that the Jayhawks manage to play on the Hawkeyes a week from Thursday. I have four tickets for a sports night out—care to join Tom and I?"

"Oh, definitely. Always interested in some legal violence. And there's a ball and a basket there someplace I imagine. A proposition you say. What do we do to earn this favor?"

"I thought you could practice your skills on a couple of punks who deserve your kind of attention. Their idea of fun, overheard at a local bar, would be to mess up the lady friend of our retired professor. We object to that sort of thing generally, and to this situation particularly. Now Tom and I intend to teach them some manners; we could use some help."

"Sounds like a good project. I'd like to think that we won't get arrested by your local sheriff though. What's the plan?"

"Our sheriff is a couch potato; can't tear himself away from evening sitcoms. I can pretty much guarantee that he won't be out writing traffic tickets, or anything else after dark. And these young punks will see four husky guys in ski masks. That seems to be the popular way to not be recognized; no need to make ourselves a future target."

Ted accepted the plan for a surprise reception of the intruders; any resistance by the targets would be met by a superior force with effective ways to deal with any nastiness. He and his karate companion Raymond would be available for activity next Friday afternoon.

Kathy was due to arrive about five Friday afternoon; Cal would be on hand at the RV to meet her, explaining a need to baby-sit the grill with burning charcoal at his place. "Scotch and soda awaits of course—right after you remind me how much you love me. And with your car parked here and a light on in the RV, any young lovers looking for a place to neck will look for another spot."

Dick and black-belt Raymond had taken positions well off the north side of Kathy's drive; Tom and Ted were secluded to the south side. The intruders would probably park their vehicle near the road; it would likely be hidden by bushes, headed out. Then a sneak approach to the RV.

Or they might drive all the way to the RV and make an abrupt entrance. Either way, the members of the welcome committee were poised in running shoes to sprint to intercept their target at the RV. Other preparations included a sensitive sound detector to activate a floodlight on their quarry at the RV. There had been an early creative session of practice on conversations similar to those heard on very old western movies.

It was about 11:30 that night when Dick's cell phone vibrated with the message from the Cattlemen's Bar that they might expect visitors soon. "Heads up guys! They're on their way. The tall one is wearing an Oilers sweatshirt; the short redhead has a dirty jacket over a white t-shirt. Can't mistake these assholes."

It was the sneak approach. The RV looked peaceful and inviting. But with the floodlight, the committee made a prompt attack with a tackle at the knees from behind and a following hammerlock and face against the ground. Resistance was predictably brief. The baseball bat was lost in the shuffle. A quick pat down revealed a .25 caliber pistol and a switchblade. No problem about identification. "So you're looking for excitement—you got it!"

The belligerent redhead complained, "What the fuck's going on? You got no reason to jump us. Mind your own business!"

"Ah, but this is our business. See, we are friends of the lady who lives here. We know what you're up to and think maybe we have a cure for that. But first, we plan to beat the shit out of you—sounds like fun doesn't it? Let's see. A bit more pressure of your dirty face on the ground and there will be a nice nosebleed—maybe a few loose teeth as well. Then we can get serious with broken fingers—both hands I think. Should be able to manage a dislocated shoulder too. Now for the knees, we'll need to find a two by four—must be one about nearby."

This provoked some struggle, but the hammerlock tightened and a bit more pressure of face on the ground started the nosebleed.

"Oh my! Look at all that blood. So sorry! Just relax—by the time we finish you won't be able to feel much of what we do with that nice sharp switchblade. No problem a'tall!"

Now there were muffled screams from one, whimpers from the other.

"Hey guys! This is going to take too much time. Can't we just go directly to step five. Then we can go party someplace."

"Hell, these babies are just going to empty their guts all over the place. Messy as hell! Why don't we just postpone this for another time. That is, if they are still in the territory."

"That's a thought. Don't want to mess up my nice new running shoes. And if they left town in a hurry, we wouldn't have to bother."

"Shit! I was just getting in the spirit of things. But these assholes aren't worth arguing about. Hate to see them just walk away though."

"Tell you what. If they was to leave here in their birthday suits, it might remind them of what's in store if they don't leave Kansas by sundown tomorrow."

"OK bozos. Strip to your shorts. There's a can of gas over there. We'll have us a bonfire of your filthy duds. Then we'll escort you out to your truck or whatever. If you don't like to leave that way, we'll just rethink our program."

"So! Do it! Shoes and all!"

"OK, there's your keys and traveling money. We'll keep this Texas license in case we need to look you up. And if we hear of any shenanigans like you had planned here, we'll find you and make vegetables out of you. Sundown tomorrow, you hear?"

So the Texas riffraff made a speedy departure. "I think we scared the living shit out of them—literally. They'll be long gone tomorrow."

"I wish I were certain of that. Logic may not work for characters like that. Think I'll camp out behind the RV with Dad's shotgun—leave the light on. If they show up again, they will be picking birdshot out of their behinds."

The committee began monitoring the Texas riffraff situation early Saturday morning. Their Texas truck was already absent from the Baker homestead at dawn. A casual patrol of the streets of Lynn and the restaurants and motels of Masonville yielded a blank. By midafternoon it became apparent that evil had fled the scene. Ted and Raymond returned to college Saturday evening. A brief call to Cal established the all clear.

Calvin responded to this news with an explanation to Kathy that vandals had been surprised at the RV by the neighborhood watch committee organized by Tom and Dick. They had been driven away, but it was apparent that the RV and its resident were vulnerable. Kathy's overnight with Cal had obviously been pleasurable and she was amenable to relocating the RV near Cal's abode. The move was scheduled for Sunday afternoon; classes at the college and university would meet as usual Monday morning. The boys had the electric and plumbing operating before they left. Cal was giving thought to construction of a wooden patio and walkway using some of the salvaged lumber of Kathy's farmhouse.

Chapter Seven

SETTLING IN

The Music Room

It was on a Monday morning in March after breakfast that Cal surveyed his 'spare room'. The wood floors rescued from Kathy's house had been sanded and refinished. The exposed rock wall at the rear of the room remained in original condition except for a casual dusting. The large picture window on the front wall effectively framed a magnificent oak tree that had miraculously escaped storm damage. There was a modest entrance door from the kitchen; the exit at the southeast corner was the substantial front door no longer needed from Kathy's house. The other concrete walls had as yet received no attention.

Cal had a partly formed inspiration to make use of the room. Kathy's piano students had remained faithful. Her Saturday morning sessions with them now took place with the old upright in the church basement. Kathy maintained the rigid discipline and lesson quality, but there was a general lack of inspiration from the absence of the old grand piano; she had frequently bemoaned its relegation to warehouse storage. If the piano could be rescued from storage and brought here, the 'spare room' could become the beginning of a 'music room'. Some careful repair of the damaged leg could be accomplished with steel reinforcement rod; the work need not be visible. Some cosmetic attention to the fracture site and 'baby' would look good as new.

Perkins promised a late Tuesday afternoon delivery if he could round up some reasonably sober muscle from the Cattlemen's Bar. Cal's feverish attention to the damaged leg was concluded just two hours before Kathy's expected arrival from University on Thursday. The carefully polished instrument emphasized the stark contrast of the bare walls. Cal was hesitant to attempt further preparation for Kathy; she would be more knowledgeable about the acoustics.

Kathy's arrival was initially celebrated with Scotch and soda in the RV. The sky was a pleasant shade of pink; best not to think about the work of the dust in scattering the blue out of the late afternoon sun. Kathy noted that the east entrance to Cal's cave was ajar. "You've been working on your 'spare room' again! I'll bet you've found a billiards table someplace to make it functional. Show me!"

The sight of her piano brought a gasp and exclamation, "It's here—my baby!" Then a rush to the keyboard and a series of chords. There were obvious tears as she returned to embrace Cal. "Oh, what a beautiful surprise! That piano is such an important part of my life. Now you have a complete woman!"

"Kathy, the room is so bare. Tell me what to do about it—pictures, carpet, furniture, whatever. I'll try to find what's needed in your ware-house. You and your 'baby' deserve a beautiful room. And I hope we can get the perfect acoustics for you."

"Yes, the concrete walls need some doing. But, you know that rock wall will work for the acoustics and it has some character. We can leave it." Then, as she stared at it, with tilt of head she added, "Cal, am I imagining things? From where I'm standing, there's some kind of organi-zation about it. It's kind of fragmented, but I can visualize the outline of something. It's huge, occupies most of the wall; if you stand close, you wouldn't notice. Look, there's this continuous line, then a break, and more. Finally, at the bottom—looks like four legs, like for an animal."

"Kathy, I believe you're right! I can imagine the outline of a buf-falo—a bison! Now, we haven't been that careful about removing the dirt. I'll just work a little more at the surface in that gap." Cal' fingernail atten-tion revealed a continuation of the outline at the top of the figure. "And look here! Take away a little dirt and I see that big buffalo head! My God! This is almost unbelievable; we'd best get one of your experts from Uni-versity to come take a look. This could well be someone's idea of a joke— or it could tell us who built this wall and lived here. A very long time ago, maybe!"

"Cal, I think there's more. Behind the buffalo, to the right—looks like some smaller animals, like maybe a small herd of deer. And over at the very edge could be a couple of trees. Big though; something like sequoias. This reminds me though; come into your kitchen. There are some odd scratches on the rock wall behind your sink. I figured it hap-pened with your plumbing work, but I wonder."

"Mustn't say that when my boys are around; they would be insulted!"

Kathy ignored him. "There, see?"

"Looks like a stick figure of a person—something out of the Sunday comics. It's small compared to our imagined buffalo—about the right proportion actually. Is it pointing to something ahead? They both seem to be on a common approach towards something, a goal of some kind. Are you trying to make some kind of storyboard out of my wall? If that's it, you can make out there's something down at the bottom too—a fish with whiskers, like a catfish!"

"And I see a small animal behind the person—might be a dog, or a fox! Think we need to look in your study. Come quick!"

In the study, Cal joked, "Kathy, I'm beginning to wonder about you. I don't see anything. End of story—or fantastic imagination."

"Maybe if we change the lighting. But I don't see any more animals or people—not even a stick figure."

"Yes, all I see is that pair of lines—a horizontal one at the bottom and a jagged one above—there just beside the fireplace. "

"Hey, that one looks an awful like Long's Peak. Look how it's above everything else and flat on top!"

"Kathy, you've had too much Scotch. I can't see it! But we need an expert to tell us if these scratches are more than our imagination. And if real, what do they mean?"

"I'll talk to Professor Sauer from Anthropology. Maybe get him to come out to take a look. And do you suppose this might be related to your irrigation ditch out front?"

Further discussion of the wall was distracted by the need to organize dinner. Cal apologized for the hamburger; "I have one of those ready made salads, some red potatoes, and frozen peas. I haven't found any interesting recipes that work well with my cash flow problems. And the wine is pretty questionable; even California wine is getting expensive. Got a bottle of southern hemisphere to try; you'll have to be the judge if it's worth drinking. Maybe this will be a bad enough disaster that you will be inspired to make out a shopping list that will be worth looking forward to. I'll do the grocery run—make this a team effort."

"Ah, yes. The way to a man's heart—thought I had solved that problem. I suppose some good food would be some insurance if you start getting bored."

"Can't picture such a scenario for another 20 years or so. But this dinner is simple enough that I won't need assistance—or any kibitzing. I'll pour the wine and you can give some further thought to the music room. Make a list of things—I'll try to find them in the warehouse this week. Some of the pictures were smashed though, sorry to say."

Actually, it was not difficult to do justice to Cal's supper. Appetites were satisfied though. Dishes were set aside for treatment at some necessary future time. Cal explained, "I've programmed the coffee to perk—or I guess it's drip these days—in the RV about now. We'll have a nice westerly view of the sunset from there. And my chocolate cookies await us there as well. Made them from scratch out of a box. They're sort of crunchy and a little dark on the bottom, but edible. Then you can tell me what to look for in the warehouse this week."

At the RV, "Found these coffee mugs with Maori designs. I seem to recall that we were assured there was no lead in those designs; fortunately nothing on the inside in any case. Drink up—and it's permitted to dunk the cookies if necessary."

"Those concrete walls in the music room—suppose you can attach those weathered boards from my old shed there? Then paint them some off white shade to brighten the room a bit. That will at least be a start on the acoustics. I'd like some of my old prints for the walls if you can find them. And some furniture—a chair or two?"

"Probably. But your desk bit the dust, sorry to say."

"No loss! I'll make do, somehow. Ah, but I have an idea to brighten things up. Suppose we could order a few of those gorgeous sheepskins from New Zealand to dress up the chairs and vicinity? Be a nice reminder of the beginning of things intimate!"

"You remembered! Thought that was a nice touch if I may say so! Yes indeed! I'll check on the Internet—or e-mail some of our New Zealand bed-and-breakfast friends for information."

Kathy's departure on Monday morning was the usual poignant experience. After the last tender kiss, Kathy said, "Smile for me Darlin'; it's only for a few days. And you're going to get busy on my music room, right?"

By midweek, Cal had the weathered lumber cut and stacked for use. Tom and Dick had been alerted for Saturday duty with promise of late afternoon steaks and beer. Only two large undamaged opera prints could be found. Kathy would have to do some research; perhaps there would be something suitable in a University storeroom.

Where Corals Lie

A Thursday afternoon university faculty meeting postponed Kathy's arrival to midday Friday. Then Cal found himself playing second fiddle to the grand piano. Not unpleasant actually. Not only was Chopin pleasant listening, but there was something comforting and reassuring about this loving domestic future. A leisurely Saturday morning of love was not to be, however. Tom and Dick arrived shortly after sunrise. While they pretended to be models of discretion, there was a certain amount of door slamming and muffled chuckles out at the RV as they made coffee.

Dick had solicited some professional advice about doing this indoor trim project and the job moved along briskly with Calvin cutting pieces to prescribed lengths and his young workers finalizing the wall structure. By mid-afternoon, the guys were sipping beer and kibitzing Cal's roller application of water-based paint. "You missed a section there—and that other board deserves a second coat." And, "Kathy if you soak Cal's old shirt in some water, you'll be able to follow his clumsy progress around the room and get all those spots on the floor!" Finally, Cal presented the roller to Dick with the insistence that he needed to start the charcoal grill for the steaks. No argument about that!

The Sunday morning inspection found a promise of future music—and a slight odor of fresh paint. Kathy insisted on a formal placement of her lover in relaxed position beside the piano; mutual sighting from the keyboard was assured. Tom and Dick were relegated to cross-legged positions against the wall. She made the formal announcement. "Calvin Carpenter, retired professor and lover extraordinary, for this special occasion I have auditioned myself these past few days to give you a lovely reminder of our recent vacation celebration of the beginning of the rest of our life together. With apologies to Janet Baker, and inspiration from her recording, I give to you 'Where Corals Lie'."

This was a careful professional rendition, but tempered by emotional dedication and loving eye contact. Calvin's eyes were a bit moist as he answered Kathy with a passionate kiss. Their audience watched and listened with reserved approval—then with enthusiastic applause. "Right on!" and "Encore".

"So—listen up! For you guys, 'Some Enchanted Evening' from South Pacific!"

Discovery

Kathy had finished kitchen chores on Monday morning. The weather was not conducive to sitting out in the sun with coffee so she wandered about Calvin's study browsing his book collection. He was off on an errand to the hardware store for some supplies. His collection of paperbacks was in sorry shape from the storm and recovery in cold storage, but readable. The few hardbacks were in slightly better condition. *The Professor* caught her eye. Not a surprise for Cal, but she wondered about his choice of biography. But the synopsis indicated it to be fiction. Still not a surprise, but she was curious about the plot. The setting of Golden West University in the Colorado foothills explained his choice. Then there were more familiar details. A principal character named Christopher Randolf Baldwin caught her eye. Why familiar? Suddenly she recalled that to be the name of the website author on climate change! But this book was fiction!

Strange! Apparently someone had adopted that name as website author, probably for reasons of anonymity. But the information seemed to be authoritative; Cal had agreed to that. Still, Cal must be familiar with *The Professor* on his shelf. Why had he not realized the website author subterfuge? Kathy was puzzled. Probably Cal was aware that one of his professional colleagues had chosen this method to inform the public of the problem of climate change and global warming. He could have explained that when they were discussing it; why not?

Uh, Oh! Cal was perfectly capable and knowledgeable enough to have done this himself! Possible yes, and did she know this man well enough to be certain that he was not keeping this secret from her? Were there other secrets in this book? She decided to borrow it to read at University. If Cal noticed it to be missing, he would have to own up to his trickery.

It was midweek when Kathy was able to snare Jim Sauer with coffee at the student union to describe the wall art at Calvin's 'cave'. "Interesting!" was his initial obligatory reaction. Then recognizing Kathy's accusing stare, he added, "You think it's more than some teenagers' graffiti? But most of the prehistoric paintings that I've seen depict natives with weapons attacking large animals. This does seem unusual. Sounds like a harmonious situation. And you say they are all facing the mountains? I wonder—there might have been a drought like we have west of here. Perhaps this is a story of the migration of some early society to the mountains. Actually, it could just as well be a record from the 1930s of abandoning the land during the dust bowl."

"I don't see any wheels of a Model T in the picture though!"

"Just a thought. And some of my students are pretty imaginative. I get this feeling they may be watching to see me get all excited about this. Tell you what—I'll come visit you at end of term and take a look-see. I suspect some imaginative fraud by one of your neighbors, but take care of it, just in case."

Kathy promised, then explained, "You must understand. It is now part of our life. We have to believe in it no matter what you find. Not scientific I know, but sort of romantic!"

Harmony

The mystery of the possible earlier inhabitants at Calvin's 'cave' prompted Kathy to wonder about their religious belief or philosophy of life. Professor Sauer had suggested the outside possibility that they may have been early practitioners of living in harmony with the land and its creatures. This was personally appealing to Kathy; the consequences of climate change made such a practice attractive, in fact necessary. There must be some way to weave her music profession into this philosophy. Of course, the inspiration and the foundation for the project obviously originated in the development of the close relationship with Calvin. His familiarity with the story of *The Professor* suggested a starting point. She would begin with the words of the Professor. And there was a mischievous need to use this as something of a dedication to his profession and to their getting to know each other's secrets.

A discussion of a musical composition with this motif to Edgar Bailey, an accomplished cellist with an adjunct appointment at University, was the beginning of weeks of creativity. He was delighted to arrange the support of his string quartette colleagues and the technical expertise of the university recording studio. The project soon had a life of its own.

It was an early copy of the DVD that Kathy brought to Calvin's cave for its first airing on a subsequent Friday evening. "Cal, our discovery of the figures on this rock wall has inspired a little production at University. I mentioned that Dr. Sauer wondered if that person and the creatures suggested an attitude of living in harmony with the natural environment. And I know that you have a similar feeling about the good simple things of living along the Raccoon with the birds and critters that call this their home. Then the violent weather made it obvious that we could not consider ourselves separate from Nature. And I have discovered that you have some deep thoughts of responsibility about that relationship. So, I had a discussion with Edgar Bailey and his friends at University about a cooperative project to express our thoughts of living in harmony with the natural laws of our planet. I'd like to show you the results of our efforts—hope it fits with your philosophy."

"Kathy, I am confident of your professional ability to create something good—and I would like very much to see your version of harmony."

"Good! I've set it to play on your machine. So sit back in your TV chair and get comfy. I'll be right here to see your reactions."

The program began with the string quartet in casual attention and Bailey making the introduction. "We are here in support of our lovely friend, Katherine Martin, who has some thoughts of life on our planet that she wishes to share with you." The camera shifted to Kathy seated at the piano playing a few bars of the introduction to "America the Beautiful". This was a stage setting not seen before by Calvin. Kathy was dressed in a gorgeous black gown clearly chosen for this special event. A conservative makeup job completed the magnificent portrait. Cal sat up abruptly exclaiming, "Oh, my God! You are one beautiful lady! How is it that I am privileged to know you?"

Kathy's voice of the opening verse of 'America the Beautiful' with piano and strings was the introduction. Then she stood with the explanation, "Kathryn Lee Bates was inspired by the beauty of the land. This led her to write in praise of the culture of its inhabitants and the potential of a great nation. But now, in these days of climate change and awareness of our errors in care of the land, let us recall the tremendous forces of the natural world that created our physical environment and our fellow creatures. Alex Cook's *Professor* described it well; let me read it to you with the cello accompaniment it deserves."

"Look at the majestic picture of our land; imagine the formation of our mountains. Perhaps a molten magma gradually lifted an overlying crust. The seas, which had covered the region for perhaps millions of years, began to roll away, leaving their accrued sediments. Molten lava bubbled up in a violent display; a few volcanoes here and there, mostly just one huge mass of cooling basalt. The atmosphere responded with violent convective winds producing electrical storms and heavy precipitation. Finally, as some of this internal heat of the earth was radiated into space, the climate cooled and snow began to replace rain in the precipitation. But the resultant seasons were not sufficient for the sun to remove the accumulating snow. It became thousands of feet thick in mid to high latitudes. The resultant glaciers carved huge U-shaped valleys, grinding the granite into boulders and gravel. Another climate change occurred and the glaciers retreated, even as they continue to do today."

"It's comforting to believe the mountains will always be here. Geologic history aside, I imagine them to have a soul that lives, with lifetimes that are eternal. They passively observe and absorb the changes in the forest and its creatures. On the other hand, the forest is relatively active. These lodgepole pines and spruce stand and watch me pass by and are entertained. Last evening as I moved quietly among the trees, I came to believe they were aware of my troubles and were whispering among themselves about me. It was reassuring to know that they would stand watch during the night and be ready to welcome the new day and me at sunrise. They have a patient overview of events. They revel in the strong winds that herald the arrival of mountain storms. They expect the clouds to refresh the soil that roots will collect nutrients. Then the sunlight will transform carbon dioxide into new layers of wood. They endure the wildfire; the forest will live again as their cones spread their seeds. Were any creatures present to observe these things? Perhaps they were not so different from us, but now their molecules exist only in the air and the sea."

Kathy resumed her version of environmental changes. "Thank you, Professor. Now today we may add observations of our changing climate and the surviving creatures of our land. We humans have multiplied and prospered; we have become an influential part of nature. But we have made mistakes in exploiting our resources. Our voracious appetite for energy has given an added dimension to climate change. Our domination of our fellow creatures in harvest of their bodies and habitat modification now cause us to give increased value to their presence. The ancient philosophy of living in harmony with the land has new meaning."

"Certain creatures have learned to adapt to our presence. The coyote, fox, and raccoon coexist with us. Robins and whooping cranes continue to visit where habitat is favorable. Deer, elk, and bison are in magnificent view where space permits. The wolf, the grizzly, the cougar contest our dominance on a level playing field; wisdom and respect will grant them continued majesty. But ants, mosquitoes, and spiders may yet survive us."

"And the Professor has given us an example to consider. He expresses admiration for that marvelous rodent, the beaver of the Rocky Mountains. He explained: "The beaver was nature's original engineer. A pair of beavers constructed a dam; they have protection from predators and a place for a lodge. Family life began. The dam would have assured a year-round water level, not only for the well being of the beaver families; now we would have excellent habitat for brook trout and all sorts of other water creatures. The water table would continue to be high and stable with lots of lush vegetation nearby. There would soon be varieties of butterflies, songbirds, and probably a watering place for deer and elk. It is a beautiful spot with a view of the Continental Divide; I would be happy to share it with the other critters. We know the beaver to be industrious and intelligent; they are not vicious competitors. The challenge is for humans to learn to live with them in harmonious coexistence."

As the cello faded away, Kathy had more personal observations. "The Professor was an educated observer. His discipline of Environmental Science governed his approach and his interpretation. He was imaginative, but he was faithful to the bottom line—the way things truly are. I too am an observer. But I am a student; I have no constraints. I find that I am one with Nature."

"I know that those of earlier societies were hard pressed to subsist on equal footings with other creatures. Still, they had the opposing thumb and better brain capacity. While they began as part of nature, they developed tools and solved problems. They began to view themselves as a superior society; Nature was there for exploitation."

"I have been part of that society. True, I did not hunt for food or gather fuel. But I served a cultural need for those whose business was the extraction of coal and minerals for my use, or the harvesting of timber for my house and furniture. And I have leisure time to think. I have the advantage over other creatures in being able to think of the future. I know that I will sometime die—hopefully not as the rodent in the talons of an eagle. But I will go. And the planet with its trees and animals will continue. I wonder—will it be a better place for my time here? Or will the magnificent structure crumble?"

"I am part of that structure. The rabbit that I see in my garden breathes the same air as I. He feels the warm sunlight as I do. And he enjoys the lettuce as least as much as I would. His life will be short. He will search for food, copulate, and make new rabbits. But he will not anticipate death or care for his environment. I am tempted to be so vain as to believe this planet, this nature, was designed for my service. I am capable of demanding that. But I know that I must share the air and the sun with the other creatures. I know harmony. Now I must use my intelligence and tools to be a responsible part of this system."

"Now my formal education has not prepared me to make more than a few token gestures to the careful use of earth's energy. I am aware that more impressive steps must be taken for the health of my planet. But I know one who has this knowledge, and the inspiration to communicate this need to society. My best duty and pleasant role will be to give him my music and my love."

"Finally, let me give voice to a strange little melody born for this occasion with support of piano and cello."

She began her song with a wistful declaration. Then her recital of these simple phrases became a hint of passion for a life in harmony with the natural world. And the conclusion faded to a hopeful benediction.

"Now for a brief time,
        we inherit this beautiful land.

I sing!

I sing with respect
        for universal forces
        that shape our world.

I sing in admiration
        for life
        of plants and trees,
        and fellow creatures.

May this harmony
        abide with us
        in all our fleeting days."

Cal's startled response to Kathy's musical introduction with 'America the Beautiful' prompted a pleasant and wondrous glance. Her voice and the care and polish in presentation did justice to the familiar song. Then the recognition of the contributions from *The Professor* began to provoke some unease. Did Kathy realize that he was guilty of adopting the Chris Baldwin character for ghost writing the website contributions on climate change? The hesitation was brief. She was observing his expression with an affectionate and sly smile. He could only nod his embarrassed acceptance. Now full attention was given to the composition.

"Kathy, you have created something really special! The Professor's soliloquy will be reborn. You have described an awakening to the true meaning of harmony with the land and its creatures. And that brief touch of your beautiful voice at the end put goose pimples on my goose pimples and a lump in my throat! You are truly wonderful." This said with a tilt of her chin and tender kiss.

He continued, "I take it that you discovered my copy of *The Professor*. And have blown my cover for the Chris Baldwin website. I've been feeling more and more guilty about that in recent weeks. At first, there was no particular reason to explain my subterfuge in doing a bit of professional communication. But then I've come to want to bare my very soul to you. And admitting this secret became an embarrassment. Please, can you forgive me?"

"Cal dearest. I've taken considerable pride and enjoyment in springing this little bit of creativity on you. It was my way of exposing your keeping a secret from me. I'm very happy with your kind of secrets! Another kiss, please!"

Lynn Journal

Calvin had made a grocery run into Lynn to replenish his supply of lettuce, celery and tomatoes, and a major investment in wild king salmon to splurge on Kathy. Then a lunch at Ben's Cafe. Jasper Barnes, editor of the Lynn Monthly Farm Journal found him there. "Cal, you'll remember I talked with you a while back about your views on this climate change thing. I was particularly impressed that your attitude meshed very well with that website presentation by Chris Baldwin. Not a surprise, I guess, to have two scientists agree. But as I recall, my neighbors here were pretty argumentative about that subject when you gave that talk at the Chamber luncheon. You might be interested to know though that they have begun to think a little more objectively about it."

"That's good to hear. My real goal was to get folks to think analytically about it—rather than just accept anything the politicians and talk show hosts say."

"You know, that's just what that fellow Chris keeps saying! Imagine that! And that subject came up the other day when I was visiting with the editor of the big city daily up north. We were at a meeting called 'Responsible Journalism'. He was thinking about getting more scientific input in his editorials. Now what I'm getting at is, I'd like you to write a guest column for the Journal on your views about climate change—not a technical scientific report, just your interpretation of what's happening. Something like that luncheon presentation—folks certainly had no trouble understanding it—gave them something solid to argue about! Now, I couldn't pay you a lot, like you were a professional journalist. But maybe an honorarium of something like a hundred dollars—you could probably take your lady friend out to dinner once or twice on that! What do you think?"

"Jasper, it's kind of you to think of me on something like that. That would really be an honor. Actually, I feel a responsibility to continue to communicate my science to the public, regardless of the cash. I'll have to think about it though—give me a few days."

"I'm sure what you have to say will be worthwhile. In fact, I suspect there may be some interest from my editor friend up north to give it a wider distribution. Can't promise that, but you deserve some token of payment for your time and effort in any case. Please give it a go—keep you off the street don't you know. I'll check with you on it next week. Now, keep that truck between the fences. See you!"

It was on Kathy's Thursday evening return that Cal asked for advice. This occurred over the celebratory Scotch; this was delayed slightly by the initial affectionate hugs and kisses—and the attention required by Sheriff. "Kathy, I've been asked to write a column about climate change for the Lynn Journal. I feel professionally obligated to do that sort of thing, but I wonder if I'm up to it. It's a big responsibility."

Kathy was elated. "Oh, Cal! That's wonderful!" This followed by another kiss with renewed passion. Then she laughed. "What do you mean—are you up to it? That website ghostwriting project was very well received. The readers were generally quite appreciative of the science, and you've demonstrated the knack of communicating. You can do it; go for it—and time for some recognition!"

"Well, you'll have to help me. Tell me what approach to take on this one. Don't want it to sound like another sermon on 'ain't it awful—and it's all your fault'!"

"Cal, you know I will. Of course, you have already convinced me of the science. Still, I can identify with the unwashed public on this. Might be able to give some advice, but you've already demonstrated your success."

"Best not jeopardize that by exposing the subterfuge and diluting the reputation of that Chris expert! And my approach will have to be quite different. I'm thinking of this as being a personal objective analysis and a somewhat philosophical prediction of the future. Like are we going to continue with our arrogant damage to the planet? Or is humanity going to recognize the scientific limitations on our future?"

"Darlin', time to do justice to your groceries. Then maybe some dedication to the release of some of your testosterone. If that inspires you, I'll let you begin your project to give enlightenment to humanity."

****

The sun was beginning to shine in the bedroom window of the 'cave' on Friday morning when Calvin brought cups of coffee to share with Kathy in bed. "Bacon and eggs and toast are on the menu when you're ready. Or may I suggest an encore of last evening's activities before we move to the kitchen?"

"Hmmm! Let me dwell on that idea while I finish this coffee."

The sun was higher and brighter when breakfast preparations got underway. Later, during the final moments of the second cup of coffee Kathy suggested, "I'll do dishes if you want to start on that composition."

"Deal! But now my sweet, how do I begin with this serious message in a way that won't turn off all those folks who are looking for entertainment? A few people are beginning to pay attention to the global warming problem but most are just concerned with the winning prospects for the Chiefs or maybe the next episode of 'The Young and The Restless'. Farmers worry about the weather of course, but can't imagine they can do anything about it."

After some thought and another sip of coffee, Kathy replied, "I know, most of the effects of climate change seem remote. That is, unless the wind just blew the house away. Now I feel personally involved!" Another sip of coffee and she continued, "But they've heard about melting glaciers for years, and floods in Bangladesh or lack of rain in the Sahara seem like old stories. These don't have much to do with the cost of tomatoes. The price of gas at the pumps is closer to the problem. But that's just a nuisance that the government has to fix so everybody can enjoy their SUVs."

Cal agreed, "Right! Now if we were about to run out of coal and oil, the problem would solve itself; doesn't look like that will happen soon enough though. Like you say, the symptoms are remote. It's an imagined problem for just a few wild-eyed scientists."

Kathy began to stack the dishes. "So, first thing is to convince folks to pay attention. Global warming is the real thing; go for it!"

****

Global Warming
A Contribution to the
Lynn Journal

"It is with considerable humility that I take this opportunity to communicate my analysis of the subject of global climate change to you. It is an obligation that comes with the inheritance of proper genes sufficient to obtain a scientific degree and that modicum of wisdom about life in general that comes with age. What I have to tell you is not new information; it is unfortunately not a sufficiently organized body of common knowledge to require universal attention and action.

There are facts about the changing climate that are incontrovertible. These are not in the class of political opinions; they are part of the present body of generally accepted science. Let me first summarize those that are not subject to the whim of politics, religion, or talk shows. Observed surface temperatures show a modest upward trend for the past century, and the rate is increasing. The world's glaciers are melting at an increasing rate, the Arctic ice is decreasing, and the permafrost is melting. Sea levels are slowly rising. All are consistent with global warming.

How could this happen? Well, you see the earth is actively supplied with energy from the sun. By virtue of the sun's extremely high surface temperature, its maximum radiation intensity is in the visible region of the spectrum. The earth, in a near circular orbit at 93 million miles from the sun, receives a fraction of this energy and loses heat only by radiation. At the earth's lower temperature, the outgoing radiation has maximum intensity beyond the visible, in the infrared. This balance of incoming solar radiation and the outgoing heat radiation determines the earth's surface temperature. But, if there were no atmosphere, the earth would be expected to have an average surface temperature not quite warm enough to encourage life as we know it.

Actually, the outgoing infrared is partially absorbed by certain molecules in the atmosphere, especially water vapor; the heat radiation resulting from the processes of absorption and re-radiation is trapped between this atmospheric blanket and the earth's surface. The planet's temperature is substantially warmer. We call this the Greenhouse Effect in analogy to the operation of structures for protection and growth of flowers and vegetables. Life on earth has benefited by substantial water vapor in the atmosphere that contributes to the greenhouse effect. This infrared absorption has yielded an average surface temperature that is about 38 degrees Fahrenheit above that for an earth with a complete absence of the greenhouse effect.

Laboratory measurements demonstrate that carbon dioxide molecules, produced in the burning of carbon containing materials, are another strong absorber of infrared radiation in the spectral region of maximum outgoing radiation from the earth. It is thus one of the most important greenhouse gases that constitute that infrared blanket.

Human populations are increasing and technology shows rapid advances. These are fueled by exploitation of our planet's resources for food, shelter—and for energy. There is a measured 1/3 increase in the carbon dioxide greenhouse gas, clearly related to the combustion of carbon containing fuels. It is this enhanced greenhouse effect that can alter the earth's radiation balance.

Those are the facts.

Scientific models of climate, consistent with the above facts, use high-speed computers to do atmospheric model calculations to be verified with present climate, and to predict future changes. These yield a consensus that increasing atmospheric concentrations of carbon dioxide are presently enhancing the global warming, and that continuing increases of this atmospheric constituent will yield further temperature increases. These predictions of the future based on the facts of science can be made with considerable confidence.

Humans have become an influential part of Nature, demonstrated in part by the facts of present climate change. We have increased the atmospheric carbon dioxide—and its lifetime in the environment is on the order of one hundred years. Consequently, we are already committed to global warming; it is now a question of how fast and how bad. Our immediate future and the extended future of our offspring depend on our response.

Already, global warming has led to disaster in some regions, such as the sub-Sahara, where failure of natural resources has led to genocide. On the other hand, a very large part of the world's population presently, or will soon, enjoy a standard of living supported by heavy energy consumption, fueled by coal and oil. This is the principal contribution to further global warming.

It has been said that climate change with global warming is an inconvenience. Yes, for those whose profits are dependent on energy production from fossil fuels, it is a minor embarrassment; major changes in their source of wealth may become necessary. For most of us, who must adapt to an increased rate of change in our environment, there will be expensive nuisances. But for those humans and other creatures whose very existence is attuned to their present environment, an inconvenience is a major understatement.

How does one cope? Denial can be a favored technique for many. For those with little inclination for careful thought, this may be supported by a misguided faith that Mother Nature is an inviolate entity. She operates independently of humans. Hopefully she will be kind. Some, who are scientifically ignorant, will scoff at the record of temperature measurements and atmospheric model predictions. Others, with selfish interests in power or profit, will trust that disaster can be postponed beyond their time. Their denial can be based on false interpretations of the scientific observations. Results can be debunked or censored for consumption by the media and gullible public. Or a corrective response can be delayed simply by a 'need for more research', a statement as sacred as motherhood.

But none of these will alter the planet's response to the natural laws of the universe. The greenhouse effect will continue to operate. Humanity's increased gluttony for cheap energy with related production of greenhouse gases makes us an influential part of Nature, with inevitable consequences."

****

It was approaching lunchtime when Cal invited Kathy to examine his introductory pages on the laptop. "How do you like them apples, Kathy? Suppose that will stir folks up—or turn them off?"

After a quick perusal, she stared thoughtfully at Cal. "You've laid the problem on them, but like you say, most of our readers aren't ready to accept or correct the mistake of overindulging in burning coal and gas to produce a luxurious standard of living." After a brief pause Kathy added, "It's the illusion that this country is so advanced and powerful that we can control Mother Nature; that sets the policy for action. No need to sacrifice our standard of living."

Cal nodded, "I wonder if our general attitude of the pursuit of wealth and power isn't contributing to this policy. Let's see what I can do with that idea."

****

"In general, our U.S. population has been blessed with copious natural resources; these have been exploited to economic advantage. A large segment of our society has become arrogant in the expectation that we deserve a high standard of living, irrespective of our role in Nature or the aspirations of other societies. We live in large comfortable houses and drive our choice of several automobiles. We can afford modern devices for communication and entertainment. Our successful leaders are CEOs of energy or automobile industries; the rest of us follow. The policies of our government are driven by economic success. The environmental attitude borders on denial of global change or human influence. The developing nations follow our example. The poor nations have no choice; they have relatively little influence in global matters.

We may extrapolate this policy into the near future. A consequence is clearly to increase the greenhouse gas concentrations; global warming will inevitably accelerate and become more intractable in the future. Our nation will have sacrificed its responsible leadership. We will have fewer friends and more enemies in other societies."

****

There was a peanut butter and jelly sandwich awaiting Cal when he finished his thoughts on the laptop. Kathy read it while finishing her sandwich. "OK, you've assessed the guilt. But what's the worst that can happen?"

Calvin grimaced, "Don't get me started there. The worst could be unbelievably bad for some of earth's creatures—extinction! Increased temperatures might be an advantage in some regions, but changes will be disruptive. Humans are most able to adapt, but at considerable cost. I can give a few examples."

But Kathy insisted that he accompany her and Sheriff on a short walk along the Raccoon to clear his mind before continuing. It was mid-afternoon when writing resumed.

****

"Suppose we choose to disregard the threat; we may distort the interpretation in order to favor personal or other special interests. Business as usual. We will rely on the power of our economy and technologic advances to cope with any imagined future problem. Our energy-dependent society might hope for technological weapons to combat the forces of nature, or to facilitate future adaptation. In the meantime, it is tempting to continue a policy of emphasis on controlling the sources of fossil energy to fuel a growing economy.

But, global warming will not make exceptions for the individual citizens of this nation. There will be increasing events of weather discomfort. Environmental degradation will become increasingly apparent with biological failures in plant and animal communities. If we stay the course, we must crank up the air conditioning, repair and guard against beach erosion, fight increasing forest fires and proliferation of damaging insects, construct elaborate systems to insure water for drought stricken farms and cities, and repair the ravages of more violent storms.

We may temporarily succeed in these endeavors by virtue of our wealth and energy use. Our wealthy nation can make adaptations to climate change. Technology is expected to develop means to sequester atmospheric carbon dioxide without sacrificing energy production. We have the capability to construct water distribution systems to revive drought regions, desalinization plants for cities, increased fire suppression for wildfires in the dry forests, coastal armoring to combat beach erosion, and relocation of coastal populations.

Each of the above requires considerable economic support. Shortfalls must be taken care of by increased taxes or increased debt. We, the polluter will be forced to pay; the war will be costly and must be financed by taxes or increased debt to other nations. Taxes promise to increase the personal pain already inflicted by climate discomforts. The mortgages of our citizens and the nation will become due. Other nations will have similar problems; we may expect general economic difficulties. And we have worsened the problem of global change. The continuing increase in energy consumption and resultant CO_2 increase will accelerate the approach to climate disaster. Ultimately it will become apparent that we have lost the battle with the natural laws of the Universe."

****

This last paragraph was well lubricated with a gift of scotch on the rocks. The first course of vegetable soup for dinner was on hold. Kathy finished her Scotch as she read his contribution. "Don't leave us like this. There must be something constructive that we can do to avoid this mess!"

Cal shook his head, "Well, we're past the point of no return. But we can stop aggravating the problem. There have already been some modest steps, but we must do much more."

There was mutual agreement to postpone further writing for later on the weekend. Kathy arranged a prompt serving of a store bought skillet fry of vegetables with suitable addition of chicken. They watched the sunset over after-dinner coffee.

It was late Saturday morning when Cal became inspired to finish off the composition. "I want you to give me your reactions to this before you head back to University. Then we can let it sit and maybe incubate for a week before sending it off to the Journal."

****

"In contrast to a 'business as usual' policy, there exists a small segment of our educated population who possess some factual knowledge of the environment and tend to think for themselves. They occasionally abandon sitcoms to become familiar with global events described on the Discovery Channel or in the literature designed for

public analysis. While the future is never entirely predictable, we might take advantage of the 'precautionary
principle'; by that, I mean that we generally endorse the
concept of insurance against future financial disasters,
and military buildup against potential attackers. The
potential for damaging climate change is no less real.
These individuals may take steps for personal energy
conservation, support of alternative solar and wind
energy production, and reforestation projects. These corrective actions are a positive beginning—with universal
adoption, one might hope for a modest reduction in the
rate of global warming.

Actually, some of our cities and states have independently adopted a more responsible attitude and corrective
action with conservation and improved energy efficiency.
And other enlightened nations have adopted the Kyoto
protocol to reduce greenhouse gas emissions. Perhaps
before human civilization begins a final descent to
extinction, some constructive efforts will succeed in coping with this human-caused climate change.

The intelligent world may come to an understanding
that we have a powerful common adversary in the natural
laws of the universe. Cooperative actions might then
occur among all civilized nations. These would include
conservation and increased efficiency in all energy consumption. Forests and oceans would be managed for
increased carbon sequestration. Transportation methods
would be modified. Public transportation would be
emphasized, supplemented by electric powered taxis.
Personal automobiles would be designed for efficiency.
Air transport would be essentially limited to private matters or large conferences; ordinary business affairs could
be conducted electronically. Electrical power would be
reengineered with a crash program of nuclear fission
reactors; nuclear warheads would be reconstituted for
peaceful power. The research and development of fusion
reactors would be given long-term emphasis. And finally,
when electrical power is nearly CO_2 free, hydrogen will
become available by electrolysis to fuel trains and
buses—perhaps even automobiles; the exhausts will con-
tain only water vapor!

But there are potential climate disasters whose time
scale is uncertain. We need to expand our use of other
alternative energy sources in the very near future. There
are relatively small-scale methods whose lead-time
should be short. Solar, wind, and bio-fuel technologies
are being rapidly developed in other societies. Our nation
should learn from their successes."

****

It was late afternoon before Kathy had the opportunity to study Cal's
work. She expressed some pessimism about the final paragraph.

"That's a tall order. It suggests a return to the idea of living in harmony with Nature. Early societies, like the one that may have lived here
on this very spot, didn't have much choice. They were hard pressed to
subsist with their rudimentary tools. Perhaps we can redirect our
advanced civilization to recognize our limitations in coping with natural
laws. How do we get there?"

Cal answered, "It will be a long haul. Hopefully, some society will
set the example to deal with the ultimate authority of the physical laws of
the Universe. Perhaps others will eventually see that their adversarial situations are relatively trivial."

Cal had some ideas for a summary that required further organization and discussion with Kathy. These continued through the cocktail
hour; dinner at Wanda's in Masonville delayed the final segment until
Sunday morning.

****

"Predictions based on human intelligence, aggression,
or compassion are uncertain. But with the prospect of
future civilization disaster from continuing climate
change, humans must recognize and cope with this
impersonal overpowering entity of natural laws. We may
call it Mother Earth, or Gaia, after the concept of a complex of positive and negative feedbacks in Nature.

However, the early interpretation of Gaia as a benevo-
lent living organism existing to guarantee human life
may be a wishful overstatement. My Gaia, the complex
of natural physical laws, must ultimately be recognized
as the supremely powerful deity. This Goddess is not
vindictive, but ruthless nevertheless. Each thought and

action by individuals, and the policies of nations must recognize and yield. Intelligent leaders must become aware of the science and assume responsibility for action. Every society, whether organized about a specific industry or resource must recognize the ultimate authority of Gaia. Every established religion must join in common respect. An enlightened and peaceful civilization may then continue to inhabit the earth."

****

Kathy read and reread this conclusion several times. Then with a loving sigh, she rewarded Cal with a tender kiss. "Cal, as good as this sounds, I suspect it will come to fruition many years after we've departed this planet. Still it's an optimistic thought for the distant future. I hope our neighbors find it an inspiring goal; humankind must get started on it soon!"

It was on Monday morning after Kathy had departed to University that Cal wrote the acknowledgements. A copy of 'Global Warming' would be delivered to Jasper first thing Tuesday morning.

****

Acknowledgements:

Principal credit must go to those scientists who developed and used the tools of research, and whose dedication and perseverance led to the understanding and acceptance of the fundamental facts of climate change.

I am also indebted to the students, colleagues, and citizen critics who have aided in the construction of the foundation for my limited scientific insight and philosophic approach.

Finally, I declare my love and admiration of my companion of these recent years of adverse climate, Kathy Martin. Her gentle nature and innate ability to perceive and appreciate the beauty around us has contributed greatly to my environmental outlook. She has aided with clear insight and careful criticism of my writing. And most importantly, the inspiration of her music and her personal support and love have been invaluable.

Chapter Eight

ADAPTATION

Buffalo Commons

It was midmorning on the first Saturday in March when the old Chevy with rattling fenders pulled in the drive. It was the familiar mode of transportation for Dick. "Amigo!," was Cal's greeting. "Ah, I see you have brought a friend. Who is this young lady?"

"Lou, this is the old codger I've been telling you about. He answers to Cal. Oh, but his lovely lady companion is another story! Kathy, Cal, I want you to meet Louise Johnson, my very special girl friend. She lives a mile or so north of my place; we've been seeing each other for a while now. In fact she's wearing a cheap engagement ring I picked up at the flea market."

It was Kathy's warm welcome, "Louise, Dick, I'm so happy for you. We were suspicious that Dick was seeing someone pretty special. And I can see that big diamond should certainly be a girl's best friend!"

Louise replied, "And it's so good to finally meet you. I've heard so much about you and Cal and this unique place. Oh, and this must be Sheriff; Dick says you are a big teddy bear."

Cal warned, "And he'll give you a kiss to remember if you're not careful."

Kathy interrupted, "Please come into our 'cave'. If Cal will heat some water for our morning green tea, I'll show you around."

Later, with tea in the sunshine, Dick elaborated. "Lou and I have been acquainted for quite a few years. Her Dad and my old man owned identical spreads just a few miles apart. In fact they used to trade corn or wheat fields regular like as insurance against hailstorms. Not a complete disaster for either that way. So you see, it's just like the old days; I'm marrying into a bigger ranch!"

Calvin queried, "So you're about to settle down on the land—in spite of the drought?"

"Well, it's true that wheat is borderline to a disaster just now. But we think we might grow enough corn and oats to feed some chickens and a couple of milk cows. And we may have a new way to make ends meet. Stanhope from East KansasPower brought up the possibility the other day. They would put up some of these big G.E. wind turbines on our ranch; pay us a few thousand every year to use our space and wind. He figures that extra power would be just enough to prevent those brownouts they've been having. We've asked the Audubon Society to determine just how far from the Raccoon the towers would have to be. Can't have those blades killing our songbirds. The birds will probably hang out near the water where there are insects and stuff for food. And hawks like to perch on towers like that to watch for gophers and things, but there aren't many of those critters out there in the open fields. Actually, the new tower design has eliminated the perching spots. We can still use the land, grow crops or graze cattle—if it ever rains again."

"Oh, I say! That could be the beginning of a general move to alternative energy. We've got to move that way soon or some kind of global warming disaster may catch up with us. And I hope it works out to keep farmers like you in operation."

"I have another idea I'd like to tell you about. We're going to bring the buffalo back to the land!"

"Say now! That might work. They were natives here for hundreds of years, at least."

"I've convinced Lou's Dad. We're each going to get a couple dozen cows and a couple of bulls and just turn them loose. They have the reputation of being easy on the prairie. They move about and don't overgraze. We'll start them out near the Raccoon where there is water and good grazing. Then, we'll try to get the native grasses to come back so the Buffs will be happy here and not head for Montana. Just hope we can all cope with the drought. Actually there's a similar project north of here. We're planning it different from the old cattle ranches though. The buffaloes will be wild—not owned by anyone. When the herd gets big enough, we'll harvest a small fraction of whatever is on our piece of land."

Louise added, "See, buffaloes don't much care about the little fences we have around here. And no way can we brand them."

"And just as well have them wild. Then when they come shit—sorry, defecate, on your lawn we can say 'not my critter!' Course, you could use those buffalo chips as renewable energy, don't you think?"

"Ah, but why don't you work on a method to capture all that methane, not nearly so messy! But, more power to you. I can see I'll have to find some way to discourage them from walking on my roof and exploring our garden though!"

"It will take a very rugged fence—I'll see to it, Cal!"

After tea and more small talk, Louise and Dick were on their way for more introductions to neighbors. Kathy's comment a few minutes later said it all, "Life goes on doesn't it, in spite of global warming!"

Garden

Then in late March, Dick arrived with tractor and equipment to prepare the soil for a small garden in the open areas along the Raccoon. Calvin optimistically ordered seed for lettuce, carrots, and string beans and planted the seed potatoes. Kathy brought three tomato plants and dozens of strawberry plants for transplanting. Dick and Louise exchanged discrete smiles. "But they ran out of tomato plants and I am willing to share strawberries with the robins if necessary. Just wait. I make a good shortcake!"

And so it was in late May that Cal and Kathy were found in early morning pulling weeds from the irrigated soil. Kathy said, "Dick, Louise, we are going to have more beans than we can handle; you'll have to help harvest—and if you can show us how to can or freeze, we'd be grateful. Maybe if your buffalo project works out we can trade some peas and beans for buffalo steaks." As Cal stood with hands braced in the small of his back, "But you know, we didn't plant any corn or pumpkins, and we have a bunch growing right in the middle of the garden."

It was Dick who suggested, "They look kind of pitiful though—not like hybrid plants. I wonder, do you suppose those have come up from seed from those people who used to live here. Let them grow and see if they turn out to be something like Indian corn—maize, and some kind of gourd. That would be real interesting. You will have Professor Sauer and his anthropologists at your door!"

And it was Kathy who gave these volunteer plants religious watering and general care. "Cal, I have the feeling that we're seeing some of the history related to those drawings on your wall. It's comforting to think we are following in the footsteps of earlier inhabitants of this spot along the Raccoon."

"Perhaps. It won't be the first time that a new settlement appeared on the site of an old civilization. Of course, their civilization was certainly modest; certainly looks like ours will be as well!"

****

A few weeks later, Calvin came down with a summer cold. Perhaps it was the early morning dew and garden exercise, or maybe a bug carried by Kathy from University, or both, but he was a miserable mess. Sitting in the sun with copious amounts of green tea and attention by Kathy seemed to help greatly. He was in no mood for visitors, but found that he could make an exception when Tom arrived on the weekend.

"Didn't realize you were laid up, Cal. Just wanted to come by to introduce my lady friend Carolyn. "Caro' is an education major at University. You'll see her frequently down here; might just set up housekeeping locally. Caro, these are my very good friends, Kathy and Calvin."

Carolyn added, "Tom has been telling me of your storm problems, and the way that he and Dick have sort of adopted you as family. I'm delighted to meet you, maybe get to be part of the family."

"Sorry I'm not at my sparkling best, but I'm happy to meet you Carolyn." Then joking, "Stand upwind of me and all will be well!"

Kathy added, "Congratulations, Tom. And Carolyn, don't let this guy get away—he's a keeper! And as for family, we have been greatly appreciative of the help we've had; we'll be delighted to have you drop by any time. Now I know Tom is looking forward to some cookies, so come into the kitchen and I'll start another pot of coffee to go with." Then, with the newly extended family clustered about the kitchen table, Kathy continued, "Now, Tom; I trust you are moving along towards that career in medicine. How's it going?"

"Slow but sure, Kathy. And let me tell you, the profession is keeping up to date. I've been hearing about seminars on preparing for malaria and other tropical diseases. With global warming, they are moving north. So watch out for those mosquitoes. Not a problem yet though. So what's with the invalid behavior, Cal? Getting old are you?"

"Oh! When I'm in hospital, you needn't bother to visit. I hear that people die in those places—don't need your happy faces to see me off!"

"Kidding, of course. You've been healthy as a horse; be up and around in no time!" Then with thoughtful expression, "Seriously though. Shit happens. If Doc Sorenson ever says you need some tests or worse, ask him to get in touch with me. I'm not the medical expert yet, but I know some who are. You and Kathy deserve the best, and I'll be disappointed if I've been left out of the loop. Carolyn and I want you around for a long time—maybe to be a godfather or something!"

Kathy overheard. Health difficulties of the future were more or less unavoidable. But not to be dwelled upon! And they were in good hands.

It's Later Than You Think!

So, from human life on the very edge, to societies obsessed with power and luxury, we continue to search for and exploit sources of cheap and convenient energy. But we foul our nests with its products. We are slow to learn, but we must begin to safeguard our planet.

The Gaia Hypothesis

It has been suggested that the processes of positive and negative feedbacks in the natural environmental laws will automatically act to preserve life on earth. Early civilizations found the weather and seasons to be out of their control, but life continued. On the whole, Mother Earth was good to them. In appreciation, and hope for pleasant tomorrows, they gave reverence to the system and called it Gaia.

Present inhabitants of our planet who are reluctant to modify behaviors in the good life have seized upon the Gaia philosophy to procrastinate. Not to worry; tomorrow's weather will take care of itself! Anyway, there are uncertainties in the atmospheric models of the feedback effects of water in cloud formation and predictions of snow albedo. And, Gaia is at work—just haven't figured out how to put her into the computer programs yet!

Or perhaps the sun will weaken slightly to compensate for our greenhouse effect. Ah, but there is no good feedback mechanism for the earth to communicate its global warming problem to the sun's nuclear furnace! And what reason do we have to believe Gaia has any particular respect for the existence of these arrogant humans? Perhaps she has in mind a modified planet with an improved form of life.

But humans have become an influential part of Nature. If Gaia were designed to include an intelligent society that could correct its problems, the present system might prevail. So, are we really that intelligent?

Yours for analytic thought,

Chris

Blue Planet Pessimism

The Blue Planet Radio was getting more pessimistic. At the conclusion of the springtime review of the mid-continent weather, Marny's comment was, "We have not been happy with these weather reports. Things seem to have gone from one extreme to another. We have asked Dr. Fitzwilliams to give us some insight on what to expect for the future. Andrew..."

"Our atmospheric models have been reasonably accurate on a global scale. It has become apparent that the Greenhouse Effect is in control; we are into a climate regime not seen in our recent history. It is difficult to make local predictions though, especially for the distant future. The winter and spring outbreaks of polar air have tapered off a bit, perhaps because of Arctic warming. But the El Nino has become especially active with flash floods in our southern tier of states. And predictions beyond the next few years, even for a global scale, are problematic. There is the potential for nasty surprises that are very uncertain in spite of predictions based on the best science. The movement of the Greenland ice sheet and the slowing of the Gulf Stream have been predicted for the distant future. There is also the possibility that the warming ocean may eventually begin a positive feedback in releasing more greenhouse gases into the atmosphere. The reported acceleration of these events is extremely worrisome."

"We are being forced into a reactive mode. The carbon dioxide, in particular, has a lifetime in the present natural environment of about 100 years. What we added yesterday will not disappear during your life—or the lives of your children. We have no choice at this point but to try to mitigate the consequences. Undeveloped countries will have a rough time doing that. But the U.S. is fortunate in having the resources to adapt. We can gradually relocate industries away from the present coastlines. We can select vegetable and fruit crops more suitable to the new climate. Our major production of wheat can be relocated to higher latitudes. But, most importantly, our energy production must be modified. Decisive action must be taken. Further increase of greenhouse gases will simply exacerbate the situation."

"Sorry to be so grim about the future situation, Marny. There may be some pleasant things develop with climate change, of course. The worrisome thing is that it is largely out of our control now. The best we can do is to try to adapt and work to prevent future disasters. I'll try to find some positive developments for our next visit. Till then..."

It's Later Than You Think!

Prof has written the following poem for me to publish
on the climate website. This is a considerable departure
from my science contributions. However the choice is to
publish it or accept the task to engage in a nighttime
snipe hunt on an island on the South Platte!

Survival Of The Fittest

Humans dominate the animal kingdom.
We are the master species—but
Whether 'tis Evolution or Intelligent Design
A change is in the wind.

Philosophies and religions we invent
Espousing man's noblest aspirations.
Then, adopted by elements of evil
For hatred, violence, and destruction.

Our sexual blessings of procreation
Continue as casual mutual bonds,
But overpopulate the planet,
And generate multitudes of abuse.

We successfully exploit our natural resources,
Liberate the energy from coal and oil,
Pollute the environment,
And increase the greenhouse gas.

Earth's trees, most spectacular of plants
shelter us from wind and rain,
nurture campfires, change CO_2 to O_2.
We harvest and fail to replant.

We theorize, research, and invent,
design and build our toys
Of transport, entertainment,
And war.

The planet will respond
To radiation imbalance,
Unwise exploitation, warfare,
And uncivilized behavior.

Were dinosaurs destroyed
By a random cosmic event?
Are humans another failed experiment
from overindulgence and greed?

Or will we reject violence and war
For wise compromise
And value a healthy habitat
Over exploitation and greed
And place ultimate respect and reliance
On our superior intellect.

Foundation For The Future

Dave Crause came down the driveway in his hybrid on Monday morning a few weeks later. "Cal, I'm setting an example for all you procrastinators. My transportation efficiency has doubled. Of course a few extra dollars from serving on the school board helps. Some of my relatives and friends must have voted for me. Hope you were one of them!"

"True, and I hear that you are actually living up to your campaign promises. More power to you!"

"Careful what you say. I'm about to ask for a favor. I seem to recall that you were critical of our science education—seemed to partly blame that on our slow response to climate change. Chickens have come home to roost. I'd like your advice on how to fix it."

"Well, my advice is free—not the very best, but it's free. I think the first thing is you need to do is catch-up. It's our young people who have to solve some of the problems. And they don't really understand the situation—it's a mystery they've inherited. Could you do a crash program to bring them up to speed?"

"Tell me Cal. We'll do it! And maybe we could use some of those Chris Baldwin websites. Could you arrange that?"

There was a long pause. Finally Cal said, "I have to wonder if you have some reason to think I might do that. Do you, Dave?"

"Cal, most everybody knows that had to be your work. It was your expertise and your personality. See, to begin with, someone missed a lesson and tried to find it in the Golden West University archives. But Golden West didn't exist! Then someone surfing the book list on amazon.com stumbles on to *The Professor* with a synopsis mentioning Golden West and Chris Baldwin. The cat's out of the bag—or was it a fox? At that point it was obvious that some scientific expert was playing a game in giving us lessons and tempting us to think about global warming. Could have been anyone in the world of course, but we immediately thought of you. And some of those lessons had personal messages for us. And of course, there was that DVD of Kathy's!"

"Well now! It's flattering to think I might have been the trigger to get you to think analytically. Can't qualify as an expert, of course. But then, you just never know what you'll find on the Internet!" But Calvin continued to avoid eye contact and gave critical examination of the garden dirt under his fingernails.

Dave ignored these evasive tactics and continued, "Anyway Cal, we're very grateful for those lessons. But there is some distress in finding there are no archives."

"Oh, sorry. I do just happen to have copies—you're welcome to them." Changing subject, "But what I really had in mind was a series of lessons from the experts at University. They understand the subject and have the communication skills. You can't expect them to travel to all your schools though. It's not consistent with conservation, much less with the cost of gas these days. TV classes are never as good as personal contacts, but I suspect the kids are really going to want this information. Having a county-wide broadcast at midmorning, say, would emphasize its importance."

"Cal, could we get you as a consultant for startup on the organization? Don't want to impose on your retirement, but you're the demonstrated expert."

"Keep me in scotch and I'll follow you anywhere. Seriously though, this is going to cost money. The kind of tax money you are getting from me won't cut it. How will you manage?"

"You remember how George Stanhope was always giving you a hard time about conserving electrical power to reduce CO_2? Well he has changed his spots. He's found the enemy in his own backyard. Not only is he working to clean up his emissions, but he has promised the cash to fund this project. I have the money, no strings attached. Go for it!"

As Dave prepared to depart, "Oh, by the way. George was called up to K.C. for an emergency power conference last week. There have been lots of complaints about these brownouts we've been having. Probably didn't affect you too much with your solar and wind. Bet you don't watch much TV either. Anyway, George has promised to tell us about it at breakfast at Ben's tomorrow morning. You might be interested. Don't expect anybody to ask for your advice; they aren't quite ready to admit to your wisdom. But if they do, hope you won't just say 'Told you so!'; not a good way to be popular!"

Another Town Meeting

The mood at Ben's was somber to the point of being ugly, especially since the electric power was slow to get the coffee started. Cal made his appearance at the end of the line as unobtrusive as possible. Stanhope arrived in an apologetic mood, wiping sweat; the A/C was just getting started.

Russell Harding, from the implement shop led the attack. "So George, what's with the power situation? Can't be money—I pay my bills, even when they keep going up."

"Well, it's a long story—but you already know most of the facts. It's basically a fuel problem. As you know, OPEC has unilaterally cut way back on the oil production. They've seen some climate changes in their parts of the world and they have taken the scientific predictions seriously. They've made their pile of money. Now they're gambling that the economies of the rest of the world can hang together long enough to keep their imports at the level to which they have become accustomed. And they are not sitting on their hands either; they have crash programs to get their solar panels and fission reactors up and running."

"Now, as you know, we have already had a problem with gasoline supplies and costs. This had to do with losing the refining capacities because of the hurricanes along the Gulf Coast. Now we've lost some of the imports as well. Now you might not think this should affect our generation of electric power; most of that is from coal or natural gas. But the machinery that get it out of the ground runs on diesel—as do the trains that move the coal to our city power plants. We're in a bind—not just from the cost but supply as well. The best we can do is rolling blackouts during peak demand and shut things down when power is not critical."

Harding objected, "But my relatives in California seem to be getting by. Maybe their power companies are being more efficient and responsible for a change. How come?"

It was Basil Jefferson who answered. "There's been increased demand out there for investments in wind and solar. They got a bit ahead of the game on that—anticipated a need and got their industries geared up. Of course most of their oil comes from Alaska too. That's still coming full bore out of the pipeline; the natives want the cash to buy their snowmobiles and such. And the folks in Alaska would drill holes in the ANWR to look for more if the greenies would let them! They don't seem to believe there's a problem with the Arctic ice or the permafrost."

Jefferson continued, "Now I've been hearing that the Washington economists say that these energy costs are just creating a temporary downturn in business. They say that a new generation of nuclear reactors will be on line in the northeast in just a few years and everything will be rosy. Why are they so lucky? Can't the State of Kansas afford to build one?"

Stanhope replied, "Actually, no! We don't have the tax base and no help from the Feds; and private companies like mine can't see that long-term investment. It's not generally known, but the Chinese have switched most of their U.S. investments into power generation here. They are the ones who are building those east coast reactors. They are still burning coal at home, but they have seen the writing on the wall. They have been into hydro lately, and have dozens of fission reactors in planning stages. Actually, it's India who is way ahead of the game on nuclear. Their greenhouse gas output is already beginning to level off."

Dave Crause interjected, "Sounds like even the developing countries are moving ahead of us on cutting back on greenhouse gas emissions!"

Stanhope nodded, then added, "We've begun to get the message finally. Some of our coal-burning plants have begun to capture the carbon dioxide for storage underground. And there's a little project in the wind, so to speak, that I want to tell you about. The powers that be up in K.C. have agreed to shell out for a wind turbine farm just north of here. Nothing so massive as nuclear of course, but the technology is well established and the costs have become competitive with coal. I've convinced them that a bunch of these G.E. turbines can make up our shortfall on coal power. The Audubon Society has agreed that it is environmentally sound. Turbines out on the Carillo ranch will be close enough to the cities to keep the transmission costs under control. I think we can be up and running in a matter of months. How do you like them apples?"

Harding exclaimed, "Good God, George. You've become a Greenie!"

"Yes, I've been giving some real thought to those Internet messages from that Chris fellow from Golden West. And you might say I've had my nose rubbed in it with these power problems. We've got to get our ass in gear before it's too late."

Farmer Jones jumped up so fast that he spilled his coffee. "Well all right! I'm convinced! Now what can I do to get on the band wagon!"

Stanhope had a suggestion. "Why don't you convince Harding to trade in one of his tractors for a stock of solar hot-water panels. Maybe I can arrange for my power company to average your present electric payments into the future after you have switched over to solar. Then you can afford to get those panels up for your hot bath water and radiant heating."

Harding was still being defensive. "George, what are you going to do when the wind stops blowing? All my lights will go out again!"

"Well, this Kansas wind will give our coal-fired systems a rest; we'll have time to get some Wyoming and Colorado coal piled up to put some muscle in our generators. And we have another option for some time in the future. These solar voltaic cells are kind of expensive now, but with mass production the price should come down. Then we can have a big acreage of solar cells working out there on that dry wheat farm west of the wind turbines. Plenty of sunshine out there; maybe backup for when the wind dies, or put it into the grid to give the coal plants a rest."

Basil Jefferson cocked his head to the side and looked up at Stanhope. "Sounds like you have a plan. You know, Miriam and I had thought about a trip to Europe until we heard that their weather had turned sour. But I've heard that investments on wind and solar are up over there—especially in Denmark and Germany. And there was some sort of international project to build a big new reactor in the south of France. There's a big push to move up the schedule on that. Fusion with a Tokamak, I think it's called. Cal, do you know anything about that?"

"Yes, as a matter of fact. Fusion power, like on the sun, has been under development for nearly fifty years. It never got much attention or financial support because of cheap oil. And like on the sun, the reacting positive charged nuclei need to be very hot. The plan is to confine the reactions with magnetic fields; works in principle, but turns out to not be easy when you increase the density of this hot plasma of charged particles! Still, the fuel source of light elements like heavy hydrogen is unlimited and the products of the reactions are not radioactive as in fission. And of course there is no emission of CO_2. So, the European Union and Japan have plans on the drawing board, in hopes things go well for the Tokomak."

Farmer Jones had the ultimate question, "Cal, we were a little slow in paying attention to your warnings about global warming. But we're into it now; anything else we can do about it?"

"Actually, we were into it long before you were born; it just wasn't so obvious. And corrections are being made in some societies. And here at home too, as George has told us—in part because we can't help ourselves. The fossil fuels are getting more expensive and oil just isn't so reliable because of problems in the middle-east.. And I think we, as a country, have to get more involved with alternative energies, like George is doing. And he has shown us that the startup time will be relatively short if we don't procrastinate. The sun is sending us thousands of times more energy than we need to live in the manner to which fossil fuels have made possible. Solar panels, wind turbines, and bio-fuels can be a pay as you go solution. Remember, we can't just shut off the problem. CO_2 in the air is at least as bad as the CFCs; once in the atmosphere it will continue to be a global warming problem for many, many years. Locking it up in trees or the ocean, or in the ground won't be easy either. Still, we have to work on the problem for the sake of our children and grandchildren." Then laughing, "Yours, that is! I have no children—that I know of." More seriously, "We could conserve of course—check the thermostat, not so much travel, fewer expensive toys. Nuclear energy, especially fusion, would solve our energy problems without the greenhouse effect."

But Calvin cautioned, "Even with these new sources though, we must give consideration and care for the environment—the land and the other creatures that live here. Obviously, we need to give more attention to the radioactive waste from fission reactors. But I'm also thinking of the renewable biomass projects like growing sugar cane to produce alternative fuel for transportation. I understand that the years of growing sugar cane in the Everglades muck has degraded the soil. Big Sugar keeps using more fertilizer; now the runoff has degraded the food for the birds—the wood storks and egrets. They are disappearing; the character of the Everglades will never be the same! But I never said it would be easy!"

Missa Gaia

Kathy returned Friday afternoon at the end of spring semester. After the customary affections, she instructed Cal to prepare the standard celebration of scotch on the rocks. "Cal, I have some wonderful news—too good to tell you by phone. University is hosting a climate conference in June. Joint sponsors are the U.S. and U.K. Met. Societies. Special recognition is to be given to two famous scientists who have worked on the Intergovernmental Panel on Climate Change. One is a prominent woman scientist from a government laboratory in the U.S.; the man is a retired Professor of Atmospheric Science in England; they will receive a modest prize of course. But in addition, there will be an annual prize in their name to be awarded to scientists for advances in reduction of greenhouse gas emissions and for adaptations to climate change. And would you believe, costs of the conference and the endowment for prizes are underwritten by North Sea Oil and a New England power company."

"It's about time, of course. And yes, it's a major turnaround for these industries to work for control of global warming. Perhaps the world is actually beginning to recognize that it's the major problem to be addressed by all."

"Cal, there's more! At the conclusion of the conference, the University Music Department has made arrangements to perform our version of 'Missa Gaia'—the 'Earth Mass'. And I have been asked to do the solo vocal!"

"Oh Kathy! Sweetheart! That's wonderful!" This followed by a bear hug and a series of kisses from ear to ear. "And I believe I recall that it includes that famous old hymn 'For the Beauty of the Earth'. When you sing that with full orchestra, the whole world will know you are in harmony with the land!"

"Cal, you will be listening from a reserved box seat with the president of the University. The concert is to be dedicated to Professor Emeritus Calvin Carpenter for recognition of work for public awareness of the supreme influence of the universal laws of nature in climate change, respectfully addressed as Gaia, in modern human life."

EPILOGUE

Influenza Epidemic

In early August, Calvin found this disturbing announcement on Blue Planet Radio. "Our weather seems to have taken a rather persistent move to drought conditions on the Great Plains with triple digit daytime temperatures. In contrast there are floods in the east and El Nino conditions on the west coast. We are fortunate to have a breather from intense hurricanes in the Gulf and Caribbean; the storm tracks have taken a northward turn in the Atlantic. This means stormy weather for England and Europe though. And that long northeast fetch behind the storms is causing lots of beach erosion in Florida. But things are really nasty in the Pacific. I hear that the western Pacific countries are being hit hard with typhoons. Dr. Fitzwilliams is here today to give us details. Andrew…"

"Yes. Although the early phase of El Nino affects our weather in California, the Pacific is still generally warm, as if the warm phase of the El Nino has not yet departed the western Pacific. This may be a consequence of a general El Nino expansion due to global warming. The result has been a series of super typhoons that have struck the south China coast. There has been extensive destruction of homes and considerable flooding. The infrastructure and health facilities have seen major disruptions. Dysentery and influenza have become rampant. Marny, I understand you have further expert information from CDC on that."

"Yes Andy. Dr. Kilpatrick is on the line from Atlanta to warn of serious consequences."

"Thank you Marny. The situation is serious. That initial flu outbreak, and the weather disruptions have permitted the virus to mutate to a more virulent strain. It has spread to Beijing where hospitals are overflowing. Now, presumably by air travel, cases are being seen in Japan and our west coast cities. It threatens to move across the U.S. in the next few weeks. The situation is comparable to the 1917 epidemic, except that our health care facilities are much better. Nevertheless, citizens are cautioned to avoid travel and to exercise precautions—wash hands frequently, for example."

****

Kathy experienced the initial abrupt symptoms of the flu just as summer classes were ending. She was promptly rushed from her classroom to the university hospital. Calvin lost communication with her on his evening call and was distraught in attempts to contact her. He frantically called Tom at University for help.

"Tom, I can't get in touch with Kathy. I'm worried. We usually check in by cell phone every evening; she says she thinks I might break a leg or something silly. Thinks I'm old and feeble; I'm not allowed to even go to town for groceries for fear of that flu. Dick and Louise bring my supplies. But today I couldn't even leave a message. Maybe something trivial, but could you check around? Call me anytime if you have info."

It was only an hour later that Tom's call came in. "Cal, she's down with the flu. She's at university hospital—getting special care. They take good care of their own. Fever is high but things are under control. She'll be there for a while though."

"Oh damn! Should have thought of that. But she seemed OK yesterday."

"Came on real sudden they say."

"Let her know I'm coming, Tom. I'll get hold of Dick to come take care of Sheriff. Then I'll get a few hours sleep and breakfast and be on my way to see her."

"Fine, Cal. I'll find you a place to stay near the hospital. Call me as you arrive. But, Cal, take care—you'll need a sterile mask, and wash hands often!"

Calvin was permitted to visit Kathy only briefly. She was weak— barely opened her eyes. Holding hands seemed to be helpful though. But he did wash his hands on leaving! It was two days later that he telephoned Dick.

"Dick how are things at home? Is Sheriff eating well?"

"Not to worry, Cal. He misses you and Kathy obviously, but he's always happy to see me. Course, I always show up with a soup bone. But how is Kathy?"

"She is recovering, but slowly. We'll be here another week or so, maybe. I'll keep you posted."

"Lou wants you to know that she has a stock of chicken soup ready. And when Kathy is up to it, Lou's buffalo stew will have her around in no time."

"Dick, I'm most obliged to you for caring for Sheriff and just generally looking after things. You and Louise are good people to have around."

"Oh, shucks, Cal! You trying to embarrass me? Anyway, this climate change thing is getting nasty. We've all got to stick together. Lou and I will always be here for you—come drought, dust, blizzards, hell, or high water!"

****

It was a few days later, after a nighttime dry thunderstorm that Dick made a careful check of Cal's computer and electronics systems. On the desk lay a folder entitled 'ZORRO'. Dick was familiar with the legend and couldn't resist a peek at the contents. There were several short discussions of topics related to global warming; they were similar to those that had appeared earlier on the Internet. Then the phrase 'Yours for analytic thought'. "Aha! Cal's work—always wondered about that!" But these particular contributions had never seen the light of day. Dick wondered, "Now the legend suggested that there was more than one Zorro—but best get Cal's permission before I make a move like that!"

So when Calvin checked in later in the week, Dick questioned him about his proposal. "Cal I apologize for snooping—couldn't resist a look at your Zorro folder. Struck me that these sections should get out there for an audience. Thought I could arrange a few blog messages—would you mind?"

Calvin could think of nothing but Kathy's well being. "Dick, no problem that you've seen that stuff. It's not a secret anymore anyway. Now my priority is to take care of Kathy. Do what you like with those topics. I have no problem with your judgment."

Dick and Louise plotted about a method to get the messages on the air without implicating Calvin. Cal's messages would be essentially unchanged, but perhaps a new Zorro could assume the publication. Dick's experiences with the degradation of weather on the farm made him inclined to be more aggressive with identification of the causes and consequences of climate change. Louise had read a novel about a mountain community noted for environmental work. They had a champion named Carlos Martinez. Such a fictional character, like Chris in fact, could not object to use of his name. But Dick had a familiarity with barnyard vocabulary that promised a more provocative analysis of society's response to the global warming problem!

****

Calvin brought Kathy home in a weakened state. Doctor's orders were for plenty of liquids and bed rest. The prescription was religiously followed for slightly over one day. She was restless. "Cal, help me in to my piano.

I need for it to help me say how much I'm grateful to be on the mend here with you." She managed to finish a Chopin Nocturne before asking for return to bed. But, passing the library she asked for a comfortable spot on the cushions with the big sheepskin. Then the order, "Cal, we need a fire." Definite improvement!

Cal was permitted a place beside Kathy after lighting the fire. "Kathy, I've never been comfortable losing you at midweek at the university. I know it's important to you for professional activity. And the salary helps us make ends meet. But you can't know how I despaired with you in hospital. If I had lost you, it would have ended my existence."

"Cal, problem solved. The state's revenue from taxes has dropped sharply; farmers are barely breaking even because of the drought. The University can no longer pay me anything beyond expenses. I shall resign—from classroom work, that is. We don't need to be apart so often."

"Kathy, that will work out just fine; I like it! We'll tighten the budget and carry on."

"Oh, I'll still give piano lessons. And the success of my DVD has encouraged Edgar and I to begin work on another. We're thinking of dedicating it to those native families coping with the climate change disasters in the Arctic."

"That sounds like an exceptional challenge! Still, I want to be your first audience!"

"Cal, I want to always be on stage for you!"

****

A few days later Calvin finally checked his computer for activities on the Internet. He quickly discovered the message purported to be from Carlos Martinez of Community.

> "Chris, perhaps you may remember that we crossed paths a few times at Golden West. I've been delighted with your messages relating to climate change and global warming. The attention was long overdue; your communication of the science fundamentals to the general public should have awakened them to the problem. Hopefully, they will begin to respond. Now I have some additional messages that I have discovered. I would like to add these; my approach will be somewhat more demanding though. Procrastination won't cut it; the problem won't just go away! So you may wish to post these as a series of blogs! Let me introduce myself and these new contributions to your readers."

Chris has been presenting the facts and fundamentals of global climate change so that you should be in a good position to analyze the observations as well as statements by self-proclaimed authorities. My name is Carlos Martinez. I live in The Community in the high country of the Front Range of the Rockies. My agenda is not so subtle. Your recent observations of the changing climate must have been comparable to the situation of the two-by-four and the mule. As the farmer said, "Got to get his attention first!" But now, my friend, it's high time to get off your ass and take action; you're part of the problem.

Now we all have our guiding principles for thought and action. Maybe it's religion, or politics, or just plain greed. But this time, it damn well better be the science! So I am going to give you a daily dose of simple science. These messages will stick to the scientific facts; hope that's not too dull for you. I may even go so far as to repeat some that I think are especially important. So pay attention! And I confess that I've been unable to resist adding some personal observations of our habit of procrastination and the unpleasant consequences of ignoring the problem. Sorry about the four-letter words.

Chris has suggested that the complex system of universal laws could be called Gaia, in reference to that early Greek idea. So Gaia has given an early demonstration of her power in responding to human modifications to her atmosphere. Our exploitation of the earth's fossil fuels has been used to generate the energy for modern civilization. Burning this fuel has increased the amount of carbon dioxide, a greenhouse gas, by a factor of $1/3$. Consequently, the earth's average global temperature is gradually increasing; this is the phenomenon of global warming. Now the global circulation has become more active, with deep outbreaks of dry continental air for some, and northward intrusions of very moist tropical air for others. Consequently our daily weather has recently exhibited extremes. This general behavior was an expected result of the initial stages of global warming. Let me elaborate on the general situation with this new title.

Understand And Endure

Planet earth is unlike our greenhouse buildings in that it is a closed system. We cannot regulate our temperature by opening the windows to space. We'd better accept the idea that things are going to get warmer. Those of you who have become kitchen experts and learned to boil water know that the water will continue to boil if the heat is on. Now, with global warming we have turned up the heat and slowly added extra energy to the atmosphere in the increased amount of water vapor. And when the vapor condenses you get it all back, perhaps in an explosive release—like a category five hurricane!

Water plays a big part in life on this planet! It is a greenhouse gas. And it makes our weather. Plants need it to grow and we enjoy it for a thirst-quencher on a hot day. So whether it is gas, liquid, or solid makes a whale of a difference!

Energy Transport And Water

The earth's atmosphere works actively to transport the excess energy from the tropics to high latitudes. At elevated temperatures, water vapor is evaporated from the oceans. The sea surface temperatures have risen about one degree Fahrenheit in the past 30 years, likely because of the enhanced greenhouse effect of increasing CO_2. Now, excess energy from the ocean surface is stored in the water vapor. The amount is large: it requires over five times as much to evaporate a certain amount of water as it would to raise its temperature from freezing to boiling. The atmospheric circulation moves this air with water vapor from tropics to higher latitudes. Then when the air cools, the vapor condenses, releasing that stored energy. And of course the other molecules of air lose some of their energy as well.

So, these processes of released energy warm the earth at high latitudes. If the radiation balance favors global warming, it seems obvious that more energy must be transported out of the tropics. With the increase in sea surface temperatures of the tropical oceans, there will be more evaporation of water into the atmosphere. And the increased energy stored in the water vapor in the tropics will be released when condensation takes place in, say, Kansas, a month or so later. Not surprisingly, the global warming increase of the energy release in water vapor condensation will result in more active weather, violent storms. But, these processes are very non-uniform over the earth's surface so the prediction of the precise local effects of a general global warming is difficult.

Water plays an extremely important part in the radiation balance and the climate of our planet. By virtue of its abundance it is an important greenhouse gas—humans are not likely to directly affect the amount. Its lifetime in the atmosphere is only about 10 days, due to the processes of evaporation and condensation. While the molecular greenhouse effect of the water vapor is unlikely to vary, the radiation balance may change with the amount of snow cover or clouds, affecting earth's reflectivity or absorption. The physical state of water—ice, liquid, or vapor—depends critically on temperature. There will be feedback effects. For example, melting ice will reduce the reflectivity for the incoming radiation; there will be a further warming—a positive feedback. Such feedback effects obviously complicate the local and global climate response. And we have these natural day-to-day variations in weather. But, riding on top of this is the relentless enhanced greenhouse effect of the CO_2.

Of course, any change in atmospheric temperature is communicated to the earth's surface. So a small increase in temperature means a tremendous increase in the total amount of available energy. And there is the obvious complication that it affects the relative amounts of ice, liquid, or vapor. And much of the planet's biochemistry of plant growth depends on temperature. Then all of us creatures must adapt. Water is crucial to all life and a change of temperature of water is critical.

But you say you are willing to make allowances for a change of a degree or so—not much difference between 77 or 78 degrees Fahrenheit. Ah, but some other place or time it may change from 32 to 33. Water goes from ice to liquid. And evaporation increases rapidly with temperature. Whether the glaciers melt or whether there is drought may depend critically on temperature. So if there are drought conditions in Kansas with crop failures, and major wildfires in Colorado, you will abandon your farm or your mountain cabin and move to Florida. Right? But Florida is being buffeted by hurricanes and that flat peninsula is in danger of being covered by rising sea levels.

HAVE ENERGY—WILL TRANSPORT!

Carlos

****

So, it is generally agreed that water plays a beneficial role for human life. We depend on it; it is our friend. I've got news for you. If you make the global warming mistake of increasing the evaporation of seawater in the tropics, you are pushing the envelope on the manner of the energy and precipitation release in mid-latitudes. Windstorms and floods are not environmentally friendly. Kansans run the danger of increased frequency and violence of destructive storms. Or perhaps the big computer in the sky will decide to move the precipitation elsewhere, like New England; the high plains of Kansas and Colorado could get very dusty. And a drink of water in the Rockies could get very expensive. Believe it!

OK, so now you know that water is not just for making lemonade! Now I have another lesson for you.

It seems that the global warming danger is not obvious to everyone—not a clear and present danger. These temperature changes so far are said by some to be just as expected from natural variations. Not to worry! Anyways, just a few degrees warmer would be nice. Warmer winters in my town will be comfortable and pleasant. Hot summers will probably make more rain from the tropical moisture, like Hawaii. Places like the Sahara are already too hot; melting a little ice in Greenland and Antarctica might be good. Nobody lives in those places anyway.

Understand And Endure

So we procrastinate even though the greenhouse effect is solid science, known and warned about for fifty years. We hear that polar bears are starving, the island of Tuvalu is awash, Katrina has flooded New Orleans. So there are a few difficulties here and there—but no problems in Kansas! Still, maybe we should pay attention to what is happening way up there in the Arctic?

Arctic Warming

The earth does not receive solar radiation uniformly. On the equinox at the equator the maximum intensity at midday is 2 calories per square centimeter per minute at the top of the atmosphere. (I know—those units are unfamiliar. But suppose that kind of power shining on the roof of your house in just one minute is concentrated to heat your bathwater; your bath will be much too hot for comfort!) On other occasions, the energy is distributed over a larger surface area; the intensity and the surface heating is less. In an extreme, at winter solstice, the Polar Regions above 67 degrees latitude (which we call the Arctic Circle) receive no direct sunlight. On the other hand, at summer solstice this region receives energy continuously; the sun does not set.

Now, with the enhanced greenhouse effect from increased CO_2, one doesn't need to be a climate genius to understand how things are changing in the Arctic! The cooling of the tundra does not occur so rapidly in winter. Temperatures in the long summer days soar; the permafrost begins to melt. The Arctic Ocean is becoming ice-free. And with less ice, there is less reflection of the incoming solar radiation; the radiation balance favors even higher temperatures. The polar bears and Alaska natives are finding themselves in dire straits with climate change. And would you believe, the science reports from polar research expeditions give detailed confirmation.

But the problem, nasty as it sounds, is not that simple. Certain societies may recognize the danger and attempt corrective action. China may begin to replace coal-burning power with hydro and nuclear to reduce greenhouse emissions. Then, the aerosol pollution will decrease; the Arctic haze will diminish. But, then the reflectivity of the incoming solar radiation will be decreased again and the radiation balance will favor even more heating of the Arctic. The problems of life in the Arctic could become even worse. And China's good deed won't remove any CO_2 from the atmosphere; it's there for another century! Uh-oh! Wonder if that was in the computer program.

IT'S TIME TO SAY GOODBYE—TO POLAR BEARS!

Carlos

****

So, folks. Now that you've had time to digest that little lesson, you should be wondering what the consequences might be for you. With the increasing radiation imbalance in the Arctic, the atmosphere won't need to transport so much energy from the tropics. Ah-ha! You say there won't be so much violent weather in mid-latitudes. Well now! I wonder how your grandmother in eastern South Dakota will cope with these climate changes.

Granny's winter climate is a bit rough, even now. In an ordinary winter, there is continuous radiation loss from the Arctic tundra through the cold dry atmosphere. Temperatures drop and a massive dome of cold air forms a large high-pressure cell. Weeks later it edges southward through Canada to our northern tier of states. South Dakota sees the beginning of the Alberta Clipper of television commentators. Conditions in the western Great Plains are cold and dry, but the collision of this cold air mass with moist air from the tropics results in an early February winter storm on the eastern plains of South Dakota. Significant condensation of dry powder snow increases the circulation about a deep low-pressure area, and blizzard conditions prevail. A flurry of e-mail messages about altered travel plans occurs. Your grandmother is sitting bundled up by the fire in the kitchen stove listening on the radio to the evening talk show with Whitey Larson, III. He is occupied with relay of personal messages from families in distress, and cautions about the dangerous weather. Granny likes to tell folks how Whitey's grandfather did the same when she was a little girl. And next morning, Granny's near neighbor checks on her to see that her furnace is operating properly, and delivers a fresh supply of firewood from the woodlot.

So in early March, Granny is reading in her winter issue of Audubon magazine about global warming; she thinks it might be nice to have Kansas winters move a bit north. Maybe not so many blizzards and the cottonwood leaves and the violets would appear a month or so earlier in the spring!

But the tropics will also be experiencing additional warming and increased evaporation from the oceans. There will necessarily be more precipitation someplace or other. Granny may find that she is living on the western edge of a continuous series of winter snowstorms in the Midwest. Her winters are getting awfully tiresome! And in spring, warmer temperatures and that additional moisture may fuel tornados, perhaps extending all the way to the Border Lakes Canoe Country in northern Minnesota.

But the predictions about local effects are uncertain. South Dakota winters could instead be mild and dry.

Then summers could be hot with high plains drought. Granny is reminiscing about the good old days of reliable South Dakota weather, and wondering if she can afford to move to Hawaii.

On to the next lesson!

There is a small minority of folks in our society who have been actively concerned about the health of our environment. We frequently call them the 'greenies'. They are well educated in the fundamentals of environmental science and able to distinguish the bullshit from scientific truth. And perhaps you are in that middle class of society that is buffeted by warnings from the greenies about global warning, or slogans about a strong economy from the business-as-usual folks. You are immersed in a climate of procrastination. The initial symptoms of global warming have been annoying, but are attributed to natural variations by the big shots in the commercial establishment. And they know about some natural process that the scientists haven't considered! Bullshit!

Let us consider this little inconvenience of the slowly rising sea level. Is this the result of global warming? All right smart-ass. I know, melting the ice floating in the polar regions won't change the sea level. The melted ice replaces only a few percent more than the displaced dense sea-water. So what is the explanation? If global warming due to the enhanced greenhouse effect of increasing CO_2 is responsible, things are bound to get worse! Warmer seawater has to expand. Glaciers are melting faster on Kilimanjaro and the mountain tops of Alaska and Peru, and the melt water from some of those glaciers goes into the ocean. If you live on the seacoast, it's high time that you give the matter some thought.

So you've noticed a bit of beach erosion and high tides moving closer to your beach house. Ah! Must be melt water from those glaciers in that National Park in Montana. But were you aware that there is also a hell of a lot of ice in Greenland and Antarctica left over from the Ice Ages. What if it takes a notion to suddenly melt?

Understand And Endure

So, you have this summer house on the beach in New Jersey or North Carolina, and a beach condo in Florida. A few more days of warm sunshine will be just fine! Your days of relaxation will be even more pleasant and the investment will prosper. But ~

The Rising Sea

Some of the effects of global warming have become apparent to you. When you made that return vacation in Alaska, the glaciers had retreated some more. And another vacation to lie on the beach on an atoll in the Pacific found the island to be smaller. And there is sea-water in the parking garage of your condo in Florida. The sea levels have risen! Now that wasn't obvious from the front porch of your farmhouse in Kansas. But careful measurements have shown that sea levels have risen by four to eight inches in the past century. The evidence on this and other environmental responses is consistent with an enhanced greenhouse effect due to increasing carbon dioxide. Now, like almost everything else, seawater expands when it is heated. And the sea surface temperatures of all the oceans have increased by about one degree Fahrenheit in the past 30 years. And the melt water from some of those glaciers and from the ice caps of Greenland and Antarctica goes into the ocean!

But why should a few more inches of water in the ocean matter anyway? The planet is mostly ocean. And the oil tankers and cruise boats operate just like always. And your view of the ocean horizon from the beach doesn't look any different. What's the problem? Ah, but beach erosion is gradually increasing. This becomes apparent during stormy periods with increased wave action. Pacific islanders, living on atolls only a few feet above sea level, find their islands shrinking as the beaches move inland. Some small islands have disappeared. Other islands have salt water bubbling up through the coral rock, killing the vegetation. Those residents have few options other than to abandon their homelands.

An extra inch or two of ocean water coming up on the Florida beach during a storm removes a lot of sand. And pretty soon the shoreline is creeping onto the golf fairway. Or that token beach dune left by the developer no longer keeps the water from lapping at the condo foundation or filling the parking garage. Your investment of 10 million bucks per acre for beachfront property is eroding!

Beachfront property along the U.S. Atlantic and Gulf coasts has been extremely vulnerable to development. Flood and water damage insurance limitations have done little to prevent beachfront construction of homes, con-

domıniums, and resort hotels. Once these are in place,
the owners and related business interests exert considerable pressure to protect and restore their beaches. Armoring these properties with groins and seawalls is a dubious
short-term protection against a category five hurricane or
a persistent series of ocean storms. Then tax supported
beach enrichment by pumping sand from offshore is
expected. This is temporary until the next storm; it also
destroys the near shore ocean life by smothering the near
shore reefs, and disrupts the nesting activities of endangered sea turtles. Well, the government can always pump
more sand onto the beach. Or barge more junk cars to
protect the shores of the Pacific atolls. And the Kansas
taxpayers can foot the bill.

The steady rise in sea level described above could be
quite inconvenient. Especially on low lying coastal areas
or coral atolls! Careful observations on the Greenland ice
cap indicate that the melt water is getting ahead of snowfall and the glaciers are moving rapidly. And at some tipping point of rising temperatures in the future, all that ice
and melt water will dump into the ocean. South Florida
and Bangladesh will disappear under the sea, every coral
atoll will be submerged, and other U. S. coastal residents
will swarm to the Rockies! Now, our five-day weather
forecasts suffer from timing, but forecasting that tipping
point is close to guesswork. Maybe we ought to take
some precautions to postpone such a disaster! How about
slowing the rate of CO_2 increase by getting more of our
energy from something besides coal and oil?

Well, now! Something to think about! With increased
global warming, such efforts to control the sea are likely
to become exorbitant and futile. Ah, you've been told
that these are just natural variations; global warming is a
fraud. Well, don't bet your beach condo on it!!

UNDERWATER PROPERTY FOR SALE!

Carlos

****

I know! For most of the natives of Kansas, a principal water sport is sailing a toy boat in the horse tank. But where are we going to put all those beach bunnies and wrinkled seniors when their condo falls into the sea?

Now the global warming aficionados have been attempting to get us to worry about the changing climate in Greenland. Wonder why! Almost nobody lives there nowadays. Even the polar bears give it a wide berth. How could anything there affect the stock market, or stop the wheat from growing in Kansas?

Understand And Endure

Greenland has a bit of ice left over from the Ice Ages. Just a bit—enough to raise sea level about 20 feet if it all melts! So what? It's cold up there and it snows a lot, especially with that extra evaporation from the tropics. Melting and snowfall will just balance, like always. But what's with all those pictures of lakes and gurgling melt-water? And NASA has that satellite that sees the ice cap losing quite a few cubic miles of ice every year! Still, one doesn't expect things to change very fast there. It will be a long time before the Vikings can move back.

Greenland's Ice

So maybe in the next century or so, if we continue increasing the CO_2, Greenland might get to be a problem. One might expect accelerated melting of the Greenland ice with global warming. Obviously, sea levels will rise at an increasing rate. Maybe you should get that South Florida beachfront condo on the market for some gullible retiree from New Jersey. Or have they been paying attention to global warming?

And the earth's arrangements for transport of heat out of the tropics will be disrupted. You see, the oceans transport about 1/3 of the excess heat from the tropics. Warm surface currents of the ocean, like the Gulf Stream, flow to higher latitudes warming the atmosphere and continents. When that dense salt water is cooled at high latitudes, it becomes even more dense and sinks to the ocean bottom to return to low latitudes. Alan Alda on the Scientific American program calls this the thermohaline circulation. Now technology hasn't been able to make water flow uphill, so there must continue to be a return current. But if there is increased melt water from Greenland, the

salt water will be diluted; its density will not be so great
and the thermohaline circulation will slow. And this will
be in addition to the increased high latitude precipitation
due to extra evaporation from the tropical oceans.

But the melting seems to already be way ahead of the
snowfall. And the Greenland glaciers are moving awfully
fast! Surprise! And some measurements have indicated
that the thermohaline circulation has already slowed
about 30% in recent years—the interpretation is uncertain, but worrisome. Slowing of this ocean transport may
be another of those unpleasant surprises—like the ozone
hole! And this little surprise isn't in the computer programs either; the tipping point is uncertain. If true
though, one may expect a disruption of the pleasant
European climate. And the atmosphere must then take up
the slack in heat transport of heat from the tropics—more
violent weather!

The problems may increase beyond what we already
see. Perhaps a reduction of the Arctic haze would be a
mistake; there would be accelerated warming. There
would be worldwide consequences. Ignoring this little
problem of the greenhouse effect could get to be a real
nuisance! The earlier mistake of adding CO_2 to the
atmosphere cannot be easily reversed.

DON'T JUST SIT THERE AND WATCH THE ICE MELT!

Carlos

Well! If Greenland does its thing, winter weather in
South Dakota may get to be even more exciting! And
Granny's summer days may not be so pleasant either.

You have probably noticed that the greenies have been in an uphill battle in seeking corrective action to the dangers of global warming. But even they have had an embarrassing surprise in the rapidity of onset of climate change, as shown by the progress to an ice-free summer Arctic and category-five hurricanes in the Gulf. Unfortunately, there was an event in our recent history that was forgotten—a lesson not learned.

Understand And Endure

We need to reexamine the stratospheric ozone problem and Chris' warning about scary things! Ah, you've been using lots of sunscreen. Good for you! But what does this have to do with global warming?

The Ozone Thing

There was a lesson to be learned. We knew much of the science of the atmosphere. But we were surprised, and may be surprised in the future, by some process not programmed in our computer. An answer to that future possibility is the 'precautionary principle'. We need to come up with the insurance premium to protect against such surprises.

Take that ozone thing for example. Somebody said that we would all go blind with extra ultraviolet if we let loose the exhaust fumes of super sonic jets into the stratosphere. Then it was spray cans with bad molecules getting into the stratosphere. Then we heard that these would just cancel each other out; not really a problem. Or if there was a problem, we should wear sunglasses; everything would be fine.

And our scientists thought they had things under control. You see, they had all the known chemistry in the computer programs to analyze the threat to the ozone. And of course, there was a lot of procrastination by industry and talk of lost jobs. The NASA scientists looking at the satellite data didn't see much loss of ozone; nothing to get excited about.

But then we had a nasty surprise. Some ground-based measurements demonstrated that there was an ozone hole over Antarctica. (NASA has since fixed their analysis.) Those chlorine compounds really did destroy ozone—all of it over Antarctica for a month or so, every springtime! That chemistry on the polar stratospheric ice clouds wasn't yet in the computer models, a completely unexpected observation.

And there was no way to turn it off! We had a big reservoir of the source of chlorine sitting in the lower atmosphere just waiting to do its dirty deed in the stratosphere sometime in the next ten or twenty years. But we had some wise people in governments around the world who agreed that we should stop making that stuff. Problem solved, sort of—you don't hear any more about it. The ozone holes will go away—in about 50 years. In the meantime, we only lost about 5% of the ozone—just a few more cases of skin cancer; not a problem unless that little spot on your nose won't heal! Actually, we were rather fortunate. If we had missed that observation for a few more years and increased that reservoir of chlorine, we would be in really deep doodoo!

So, what is the lesson to be learned for global warming? We've got the increase of CO_2 in our computers to predict the gradual temperature increase. Let's just sit back and think about how our children might fix the problem. But have we properly programmed for nasty surprises like melting of the Greenland and Antarctic icecaps, or feedback of carbon dioxide and methane from the melted permafrost or the warming ocean? And like the CFCs, the CO_2 has a long lifetime in the environment—something like a century. We stopped making the CFCs, but getting energy without burning stuff that contains carbon is more complicated. BUT ACT NOW! Our descendants could have some nasty surprises to contend with, and would be destined to live with a degraded climate for a very long time!

THIS BAD MISTAKE MAY BE FOREVER!

Carlos

****

Folks generally got pretty upset by the prospects of
ozone depletion and increased ultraviolet. The effects on
the ground of changes in the stratospheric ozone were
going to be truly global. The ultraviolet would become
more intense for every nation and for every individual.
The related increase in skin cancer would endanger every
person; children in every family would be exposed to its
cumulative effects. Procrastination was rampant though
until the chlorine chemistry above Antarctica became
definitive. Spray cans with CFCs suddenly became
taboo. Leaders in every nation scrambled to a consensus
to ban the offending chemical. We did good, but we
didn't go back to square one. Those CFCs didn't go back
in the bottle. They will live in the atmosphere for decades
and the ozone holes will remain until Nature washes the
chlorine out of the atmosphere.

So you've decided to ignore the global warming problems of the polar bears and Alaskan natives, the islanders
in the South Pacific, and the brown-skinned folks in
Bangladesh; no skin off your nose! Climate change
might actually be good for you in Kansas. It's not really a
uniform global threat like extra ultraviolet. On the other
hand, the computer calculations say the average global
temperatures will continue to increase; maybe even faster
in eastern Kansas. And if it doesn't turn out the way
you'd like, tough shit! You can't go back to square one;
it's going to continue that way for a hundred years or so!

Understand And Endure

Time to find a solution! Early humans and many societies today, manage to subsist only with the exploitation of convenient sources of energy, such as wood from the forest, or heat from the sun. And much of human society today prospers with the availability of cheap energy by burning coal and oil. Only recently has the feedback to nature become a recognized problem. Now, do you suppose we have enough intelligence and ingenuity to protect our planet?

Carbon-Free Power

The birth of nuclear power may be traced to the design and operation of the 'atomic pile' in the squash courts at the University of Chicago by the Italian physicist Enrico Fermi. This was a clear demonstration of the nuclear chain reaction initiated by the neutrons from the fission of uranium nuclei. The Manhattan Project to build a nuclear fission bomb followed in response to worry that Nazi Germany might perfect such a weapon. With this success, many scientists promoted the possibility of nuclear power from chain reactions in a controlled fission reactor. Uranium ore is available in numerous locations; refinement procedures to increase the efficiency of the chain reaction, and safety procedures to absorb runaway neutrons, are well known.

Safety considerations and disposal of radioactive waste have been dominant public concerns; the alternative of cheap fossil fuels has led to the neglect of this energy source. Nuclear energy has now become more attractive in the face of global warming and increasing cost and vulnerability of oil supplies. Safe fission reactor operation has been demonstrated in France; it is that country's principal power source. Nuclear power generates no CO_2, the greenhouse gas increasing from burning of fossil fuels. Ah, you're still frightened by those invisible rays from nuclear waste, and politicians who can't pronounce 'nuclear' are not reassuring.

Also, the release of nuclear energy in the fusion of light nuclei is the energy source of our sun. The fusion process requires the very close proximity of the light nuclei in order for the short-range nuclear force to form a more massive nucleus with the release of energy. The electrostatic repulsion of the positively charged nuclei must be overcome. This can occur in a gas of light atoms only with very high temperatures. The high temperature gas on the sun is confined by gravity. An alternative of magnetic confinement of charged particles has been demonstrated in the laboratory, but confinement of dense plasmas of charged particles is more difficult. Still, controlled fusion remains a viable energy source. There is an essentially limitless supply of light nuclei. The fusion products are not radioactive. CO_2 is not produced.

The energy needs for earth's increasing population are great. And there are some, ignorant of the science, who say global warming is a fraud. But, the dangers of global warming with increased CO_2 from burning of fossil fuels should be obvious to those of you who have persisted with these science lessons. There are alternatives. The technologic capability of nuclear fission power has been demonstrated. Nuclear fusion power remains a possibility for limitless nonpolluting power. However, the time scale for bringing these energy sources on line may not be short enough to avoid some nasty surprise not yet in the computer models.

On the other hand, in solar power we have available many thousand times our present power consumption. We need to develop the renewable energy sources, such as wind, solar, and biomass to provide energy at a rate compatible with our power requirements. So, the wind doesn't always blow and the sun doesn't always shine. But put it into the grid when available, or devise a way to store it. And if we store even a small fraction of this energy over a long period time in biomass or batteries, it can be available for peak power. The time scale for this development should be relatively short; the CO_2 rate of increase might be quickly brought under some semblance of control. The backup power grid remains essential but would require a much more modest combustion of fossil fuels. An eventual successful implementation of nuclear fusion power would likely solve humanity's future needs.

VOTE TODAY FOR CARBON-FREE ENERGY
FOR TOMORROW!

Carlos

****

Once upon a time there were dreams of peaceful uses of nuclear energy. These are still largely dreams. Rather, the atmosphere of international distrust has prompted a general effort to use it as a violent deterrent to economic or military pressures. This state of fear has extended to a worry about accidental disasters from any constructive use of nuclear energy. But these things can be controlled if we put our best minds to the task. Far better than letting climate change proceed beyond our control; we have been warned that can happen!

Alternatively, it would make sense to require all new building construction to make use of solar power, perhaps instead of that extra bed and bath. But solar panels on every home aren't going to happen without some financial help. Only a few homeowners can afford to shell out thousands for retrofitting. Government guaranteed loans might help. But it's the business of the electric power industry. There should be no new investments in coal-burning power plants except for those that will immediately operate with carbon dioxide capture and storage. Rather, how about community solar power facilities. And if lots of foot-dragging, let's have some community-owned solar power? Get on with it!

Now I am aware that there are many of today's citizens who will not have read or benefited by these messages. Some are like the dog I had. If there were something beyond his understanding, he would not look at it! But no one can escape the responsibility for care of the environment. We have become part of Nature and ignorance of her problems is no excuse. Still, there are those who follow, without thought, leaders who have some other agenda. I've said it before. "They damn well better pay attention to the science!" I don't mean you of course. But, were you listening when the little old professor gave his discussion on global warming at the library?

Understand And Endure

Science is viewed by some of you as the mysterious stuff that people with thick glasses, wearing white coats do while the rest of us carry on with the necessary chores of life. They work on important things like new pills for weight loss, the real significance of *pi,* or the structure of black holes and dark matter. About every six months they discover a new virus and spend our tax money watching it mutate. Or maybe science is just useful for building a better mousetrap, or designing a rocket ship to send some rich bastard to Mars? It's a whole other world that we avoided in public school. It's an unattractive nuisance that interferes with progress. Perhaps we could deal with it, if we just understood it!

Science

Science may be defined as that body of ideas or theories that can be tested by experiments or observations. The scientific method may begin with an observation that leads to a theory. This in turn must be tested; if the test fails in any respect, the theory must be modified. Policies or beliefs that are not testable are not science. A body of information that is dominated by governmental edict or religious dogma departs from scientific method. Such a procedure distorts the interpretation of observations and avoids the ultimate experimental test. The final authority of scientific validity must remain in the hands of individuals who adhere to the scientific method. Society in general must beware of pronouncements on climate change or any other body of science that are often made by other individuals who have some separate agenda.

There are individuals in our society who lack the education to make an objective assessment of the content of scientific reports. They are naturally suspicious of pronouncements by scientists who lack household names; these are conveniently categorized as the 'mad scientists' found on late night television. It is the public education system with administrators and teachers who have short-changed science education that originated this problem. And voting age adults who continue to be enslaved by TV sitcoms rather than Nova or Scientific American or the Science channel, (or this website!) ignore the problem.

Others, whose education should have made them more knowledgeable, have suggested that any scientific research is automatically skewed to favor the results desired by the funding individual or agency. Support from any group whose name includes the words environment or climate would be guaranteed to yield scary predictions. Perhaps it is assumed that questionable research results are simply reported on the Internet, Tom Friedman's 'river of untreated sewage'. But the research published in reputable scientific journals is invariably refereed by teams of experts who check the logic, the procedures, the arithmetic, even the punctuation!

A society's leaders may subscribe to certain policies of economics, diplomacy, or warfare. But they ignore scientific truth at the peril of our entire civilization!

CHECK Houghton's *GLOBAL WARMING* for the real thing!

Carlos

****

So, how did we get in this mess? Let's blame it on the government, or better yet, the Chinese. And there were lots of big shots out there who wouldn't listen to the science. There is that active aggressive world of individuals of capital and finance who have learned well the street rules of their society. Although cunning and resourceful, their intellect is so corrupted that they are unable to wonder at the values of an existence sensitive to the environment they have inherited. Their behavior parallels that of the dinosaur who followed an unwavering thoughtless path to extinction. The dinosaur had little control over his environment. The only excuse for many in today's leadership is stupidity! The difficulty of others to cope with climate change is unfortunately the cost of doing business. But even this powerful group will not be immune to the global warming effects. Beach houses will be damaged or obliterated. Waterfront condos will preside over eroded beaches and flooded golf fairways. Mountain mansions will be surrounded by blackened forests. Texas ranches will revert to scrub and sand.

Individuals at any level of society may be excused if they question predictions of the future. Our knowledge of the environmental science is never complete. Still, we do understand the physics of electromagnetic radiation. And there is no question of the validity of the greenhouse effect. Measurements of CO_2, a potent greenhouse gas, and its increases from burning of fossil fuels have been accurately documented. And the database of the observations of global temperature increase is well established. Our current computerized atmospheric models of the earth's climate are consistent with these facts. Individuals who use their god-given intelligence should realize that this science is far more reliable than the wishful thinking about the stock market.

RECAPITULATION AND CHALLENGE

Our planet is telling us of its state of health through the research of numerous scientists studying the atmosphere, glaciers, and the oceans. These research facts are not subject to stock market trends or military campaigns. And Earth will respond to these facts through well-established scientific laws. She may no longer endure the exploitation, the greed, and the arrogance of those societies that lack the collective intelligence and morality to safeguard the existence of themselves and their descendants.

Hello. Calvin Carpenter here. This blog contribution will summarize the works of Chris Baldwin and Carlos Martinez to expose readers of these messages to the fundamentals and consequences of climate change. I trust you have been enlightened and perhaps even entertained by their efforts. I believe that moderately intelligent individuals will have understood the problems of climate change and its dangers. I have independently presented my analysis and outlined some solutions to global warming in the Lynn Journal. I hope that one does not need to be a rocket scientist to understand what must be done. With this information, earth's citizens must arrive at a consensus that global warming is an urgent threat and demand action. Individuals must settle for McKibben's *Deeper Shade of Green* and strive for energy efficiency and conservation. BUT, IT'S THE CARBON DIOXIDE, STUPID! Mankind must make the major shift away from energy production from fossil fuels.

To accomplish this, there must be a general search for new leadership. These leaders must have intelligence, education, and have access to regulatory and financial resources. Actually, a few conscientious leaders in the governments of our cities and states have begun to make a priority of greenhouse gas control. And when a national leader calls for international cooperation, others must heed the call. Some dedicated and courageous leaders in auto and aircraft industries have independently made a commitment to reduce emissions. The giants of oil and electric power must also continue to empower new leaders who will make this shift in emphasis. There is a moral imperative to leave the earth's environment suitable for our descendants. And the handwriting on the wall is warning us of major inconveniences in tomorrow's climate if we persist in increasing the greenhouse gases. We must make carbon-free and carbon-neutral technologies our energy solutions for the future.

Respectfully yours,

Professor Emeritus Calvin Carpenter

APPENDIX

On June 21, the summer solstice, the midday sun at 23 degrees north latitude 100 miles off the coast of West Africa deposits electromagnetic energy on the atmosphere above the ocean at 2.0 calories per minute per square centimeter. (That's 1.4 kilowatts per square meter—about the power of a hairdryer on about 10 square feet.) The total daytime power incident on the earth's upper atmosphere is about 1.8×10^{14} kilowatts. This is about 14,000 times the total electrical power generated on earth. This solar power is spread very unequally over the earth's spherical surface; none arrives at night. The maximum intensity is in that part of the spectrum most sensitive to the human eye. Beyond this region of visible solar radiation is additional power of comparable magnitude. Most of this beyond the visible red occurs in the spectrum called infrared and radio frequencies. More energetic packets of energy arrive in the spectrum beyond blue, called ultraviolet. All that solar radiation arrives at the tenuous atmosphere above, say 160 kilometers (100 miles), at a speed of 3×10^5 km/sec (186,000 miles /sec). About eight minutes travel time!

Some of the radiation is reflected back into space; at least half of the visible radiation penetrates all the way to the land and the ocean before being absorbed. The Earth is warmed; the spectral distribution of radiated energy from such a warmed object has a maximum in the infrared, determined simply by its temperature. We expect the heated planet to come into radiation balance with the sunlight, since there is no other way to gain or lose energy. Now, in the absence of our atmosphere, the average equilibrium temperature of the earth's surface would be about 21 degrees Celsius lower than is actually the case. But a large amount of the earth's heat radiation is trapped below the atmosphere due to its infrared absorption by water vapor, methane, carbon dioxide and a variety of other complex molecules. This warm blanket keeps the earth's surface at a temperature suitable for life as we know it. It may be seen that the earth's atmosphere behaves like the glass in the roof of a greenhouse, hence the term greenhouse effect. While the maximum intensity of that radiation continuum from the sun at 6000 degrees Kelvin is at about 0.5 microns in the visible, the broadband heat radiation from the earth's surface at, say, 300 degrees Kelvin (27 degrees Celsius, about 81 degrees Fahrenheit) is at about 10 microns in the infrared. The radiation from the sun is beyond our control, but the absorption of the earth's infrared by the greenhouse gases of methane, carbon dioxide, and others has changed due to our industrial activity. In particular, the amount of carbon dioxide in the atmosphere has increased by about 1/3 since the industrial revolution, due to burning of fossil fuels. It is at its highest level for the past 600,000 years. Its enhancement of the greenhouse effect is believed responsible for the recent 0.6 degree Celsius average global temperature increase.

In summary, nearly all the visible radiation from the sun is transmitted by the atmosphere; much of the infrared radiation reemitted from the earth is absorbed in the atmosphere by water, methane, and carbon dioxide. The atmosphere behaves as a warm blanket, trapping much of the infrared from the earth in what we call the greenhouse effect. The earth's plants and its creatures have evolved in harmony with this radiation balance. Global warming from the enhanced greenhouse effect departs from this harmonious balance; it is an extremely serious environmental problem.

SUPPORTING LITERATURE

Anthes, R. A., et.al., "Hurricanes and Global Warming—Potential Linkages and Consequences", Bull. Amer. Met. Soc. 87, # 5 623–628 (2006).

Behrenfeld, M.J., et. al., "Climate-driven Trends in Contemporary Ocean Productivity", Nature, 444, 752–755 (2006)

Bryden, H.L., et. al., "Slowing of the Atlantic Meridional Overturning Circulation at 25° North", Nature, 438, 655–657, (2005).

Clark, Peter A, The Role of the Thermohaline Circulation in Abrupt Climate Change.", Nature 415, 863, (2002).

Darack, Ed, "Sky Power", Weatherwise, 58 #6 30–35 (2005).

Doney, Scott C., "The Dangers of Ocean Acidification", Scientific American, 294 #3 58–63, (2006).

Hansen, James, "Annual Mean Global Temperature Change", APS News vol 15 #4, p4, Figure 1.(2006).

Hansen, Kirby, "World Production of Carbon in CO_2" (From Keating and Rotty, 1973) Weatherwise, 58, #6 15 (2005).

Kerr, Richard A., "A Worrying Trend of Less Ice, Higher Seas", Science 311, 1698–1701, (2006).

Laskin, David, "El Nino Forever", Weatherwise, 58 # 6 48-51 (2005).

Lawler, Andrew and Nick Cobbing, "Photo Essay/On Thin Ice", Audubon 108,67–72, (2006).

Levi, Barbara Gross, "Is There a Slowing in the Atlantic Ocean's Overturning Circulation?", Physics Today 59 #4 26–28 (2006).

Mann, M. E. and K.A. Emmanuel, "Atlantic Hurricane Trends Linked to Climate Change", EOS Transactions AGU, 87, 233–244, (2006).

McKibben, Bill, "A Deeper Shade of Green", National Geographic Vol 210 #2, p32, (2006).

North, Gerald R., et al , "Surface Temperature Reconstruction for the last 2000 years, National Academy of Sciences Report to Congress on Global Warming", (2006)(http: \\ fermat.nap. edu\catalog\11676.html)

Overpeck, J.T., et. al., "Arctic System on Trajectory to New, Seasonally Ice-Free State", EOS Transactions AGU 86 309 (2005).

Overpeck, Jonathan, et. al., "Paleoclimate Evidence for Future Ice-sheet Instability and Rapid Sea-level Rise", Science 311, 1747–1750, (2006).

Webster, P.J., et. al., "Changes in Tropical Cyclone Number, Duration, and Intensity in a Warming Environment", Science 309 1844–1846, (2005).

SUGGESTED ADDITIONAL READING

Global Warming, The Complete Briefing, John Houghton, Cambridge University Press, (2004)

Global Warming, a Very Short Introduction, Mark Maslin, Oxford University Press, (2004)

Energy's Future Beyond Carbon, Scientific American Special Issue, September 2006, Vol 295, #3.

Climate Change 2007: The Physical Science Basis, Intergovernmental Panel on Climate Change (IPCC). http://ipcc-wg1.ucar.edu/wg1/docs/WG1AR4_SPM_PlenaryApproved.pdf

Confronting Climate Change: Avoiding the Unmanageable and Managing the Unavoidable, www.confrontingclimatechange.org, (Excutive Summary)

****

Earth in the Balance, Al Gore, Plume (Penguin Books) (1993)

High Tides , Mark Lynas, London: Flamingo, (2004)

Red Sky at Morning, James Gustave Speth, Yale University Press, (2004)

Climate, George Ochoa, Jennifer Hoffman, and Tina Tin, Rodale Books, (2005)

The Weather Makers, Tim Flannery, Atlantic Monthly Press, (2005)

The Winds of Change, Eugene Linden, Simon & Schuster, (2006)

The Revenge of Gaia: Earth's Climate Crisis and the Fate of Humanity, James Lovelock, Basic Books, (2006)

VIDEOS

The Dimming of the Sun, Nova Video
(http://www.pbs.org/wgbh/nova/sun/)

The Power of the Sun, Walter Kohn and Alan Heeger, Nobel Laureates, University of California (http://powerofthesun.ucsb.edu/)

An Inconvenient Truth, Al Gore (Documentary Film) (http://www. climatecrisis.net/)

END

Printed in the United States
85955LV00006B/123/A